GALAXY OF SCENTS

THE ANCIENT ART OF PERFUME MAKING

"Just for the Smell of it"

HEINRICH KHUNRANTH

Dr. Mary Lynne

ISBN 1-56459-458-0

KESSINGER PUBLISHING'S
RARE MYSTICAL REPRINTS

THOUSANDS OF SCARCE BOOKS
ON THESE AND OTHER SUBJECTS:

Freemasonry * Akashic * Alchemy * Alternative Health * Ancient Civilizations * Anthroposophy * Astrology * Astronomy * Aura * Bible Study * Cabalah * Cartomancy * Chakras * Clairvoyance * Comparative Religions * Divination * Druids * Eastern Thought * Egyptology * Esoterism * Essenes * Etheric * ESP * Gnosticism * Great White Brotherhood * Hermetics * Kabalah * Karma * Knights Templar * Kundalini * Magic * Meditation * Mediumship * Mesmerism * Metaphysics * Mithraism * Mystery Schools * Mysticism * Mythology * Numerology * Occultism * Palmistry * Pantheism * Parapsychology * Philosophy * Prosperity * Psychokinesis * Psychology * Pyramids * Qabalah * Reincarnation * Rosicrucian * Sacred Geometry * Secret Rituals * Secret Societies * Spiritism * Symbolism * Tarot * Telepathy * Theosophy * Transcendentalism * Upanishads * Vedanta * Wisdom * Yoga * *Plus Much More!*

DOWNLOAD A FREE CATALOG AT:
www.kessinger.net

OR EMAIL US AT:
books@kessinger.net

Special Notice

In the book, Dr. Mary Lynne expressed her frustration trying to find someone to perpetuate her work. We are happy to report that toward the end of her life she found such a person in Fallie (Pete) Bennett. She entrusted her secrets and laboratory techniques to him in full confidence. Mr. Bennett has faithfully continued the work in her little laboratory. Most of the perfumes mentioned in the book are still available in limited quantities. For further information, please contact:

Mr. Fallie Bennett
5497 Newton Falls Rd.
Ravenna, Ohio 44266

The following article is reproduced with the permission of the *Akron Beacon Journal*. The original story was published on October 20, 1968 in the *Beacon Magazine*. We gratefully thank the *Beacon Journal* for allowing us to reprint in its entirety herein.

Perfume making is a man's art . . . . but not for Ravenna's Mary Lynne Weenink. Story inside.

Akron Beacon Journal

October 20, 1968

HER HOBBY IS 'JUST FOR

One night in 1936, Mary Lynne Weenink couldn't sleep. She reached for her Bible, thinking it might soothe her. By chance she turned to Exodus 30:34, 35 and read:

". . . Take unto thee sweet spices, stacte, and onycha, and galbanum; these sweet spices with pure frankincense . . . and thou shalt make it a perfume . . ."

The formula intrigued Mary Lynne's scientific mind. She wondered if the perfume could be made and soon found most of the ingredients no longer exist. So she never made the biblical perfume.

But the idea of making perfume fascinated this tall, stately Portage Countian. She made some and has been doing it ever since. She's one of the few woman perfumists in the world, she believes. Just as chefs dominate the great kitchens, perfume-making is a man's art.

"PERFUME-MAKING all through history has been mysterious, enticing," Mary Lynne says. "It's something everyone knows something about; yet really knows nothing."

Filled with the history of perfumery, Mary Lynne can talk on and on.

"Priests in ancient Egypt discovered perfume and hoarded the secrets. But perfume-making spread through the Near East. Crusaders discovered exotic perfumes and brought flasks full back to the women of Europe," she says.

Mrs. Weenink describes Queen Elizabeth's effort to make perfume. She sent a man to the Middle East to glean the formulas but he wasn't successful.

"The only perfume he turned out for his queen was a rancid-smelling liquid made of dog fat and sour apples. The queen made members of her court wear it anyway."

MARY LYNNE'S "lab" is her basement. It's filled with various shaped

First step in making perfume is pulverizing the seed, bean or leaves. Flower petals, such as roses, are not crushed. The basic scent concentrate is then mixed with an enzyme, placed in a jar to "grow," much like flour and yeast grow into bread dough. Mary Lynne is pulverizing an orchid seed with mortar and pestle.

THE SMELL OF IT

Our Cover Story

The powder resulting from the first step is placed in a four-ounce bottle along with a digestive enzyme and seven drops from a previous batch of the same scent to serve as a starter. During six to eight months of aging, the bottle will fill. The animal fixative is added to stop growth. Thirty days later, it is ready to be bottled.

vials, bottles, tubes and mixing equipment.

Perfume is made by extracting juices from flowers, seeds, herbs, roots, leaves, trees or shrubs to provide scent. This is then worked into the basic concentrate which goes through an aging process. Finally an animal fixative is added, ambergris, musk, civet or something else.

"When I want a perfume as wild as a March hare, I use a little ambergris, a bit of musk and a larger proportion of civet," she explains.

"If you want a happy perfume," she advises, "use flowers that bloom after dark and are controlled by the moon. They are more tranquil in fragrance, too.

"Some petals, like roses, have to be picked when the dew is on them or they give perfume a depressing odor.

"If you like an invigorating perfume, one that is strong and exciting, use flowers that bloom in the daytime."

FOR MARY LYNNE, the loveliest, sweetest and most lasting fragrances come from seeds. Why? "The enzyme of life is in the seeds."

Perfumes made from resins — juices from stems and from trunks of trees — have a cooling, outdoor, woodsy effect . . . and a calming influence, the perfume maker declares.

Herbs, on the other hand, can function like drugs when made into perfume, Mary Lynne says. Some act as tranquilizers; some tend to excite and some even are inclined to release inhibitions.

"I've made a perfume 'Persian Nights' of some 59 different herbs which I'm sure the ancient Persians used in their harems," Mary Lynne says.

SHE'S AGAINST some of the colognes, after shave lotions and other scents that are big for men these days.

"They should never use an effeminate scent. No florals and the like. For men it should be citrus, leather, burnt embers and herbs."

She didn't say what these perfumes would do for men.

"But I created a scent for my husband, George, that has all his qualities. When he walks into a room, everyone remarks on the scent. It is really George."

She predicts men's perfume will be-

Continued on page 26

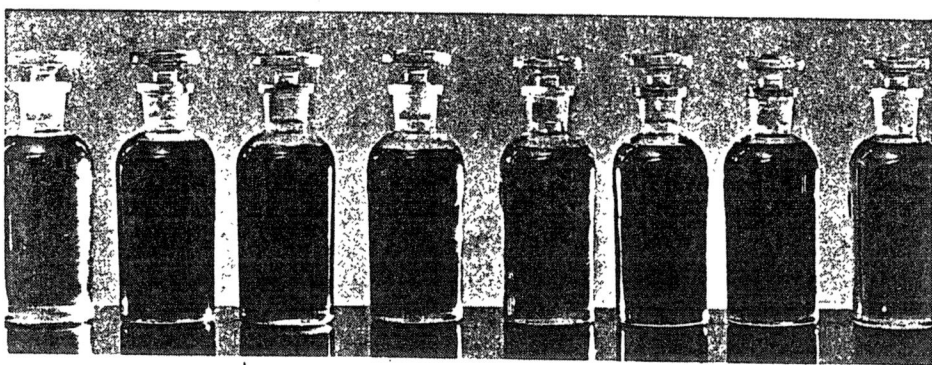

As the various scents age, they take on varied and beautiful colors.

Akron Beacon Journal, October 20, 1968

'JUST FOR THE SMELL OF IT'

Continued from page 25

come really big business, almost as great as women's.

"I MAKE PERFUME for some of the top businessmen in the country, although I do it as a hobby. I do not sell commercially, but share my perfumes with others at cost only.

"Not long ago I turned down an offer to create and manufacture perfumes

With her breath, Mary Lynne warms a gold funnel before starting to bottle the perfume. Unless the funnel is warmed, the perfume will not run into the small bottles.

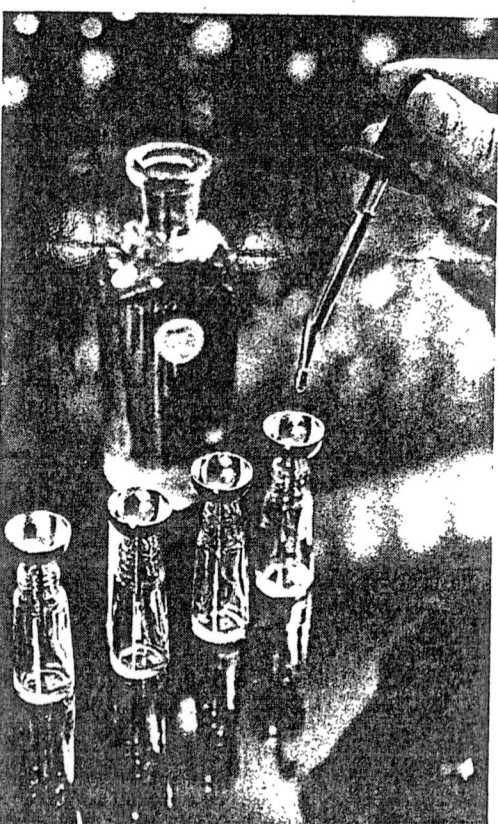

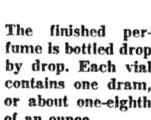

The finished perfume is bottled drop by drop. Each vial contains one dram, or about one-eighth of an ounce.

exclusively for one of the top stores in New York. The figure was dazzling but I just didn't want to get into business.

"Why should I make perfume for those who have everything? I make perfume for disillusioned housewives, working girls on a strict budget, and professional women. It's their one touch of luxury.

"It costs me about $4 a dram to make. That's what I sell it for. If I went commercial, the perfume would sell for about $24 a dram."

MARY LYNNE and her husband live in a retirement home at Silver Spur Ranch, 5497 Newton Falls rd., about three miles from Ravenna.

But it's a far cry from retirement for Mary Lynne, who formerly worked in cosmetic therapy in Cleveland, and her husband, George, who manufactured and sold vitamins and minerals.

"I'm so busy now making speeches about perfume and making the perfume I don't have a minute to myself," Mary Lynne says.

She is even writing a book, a history of perfume, which she will call "Just For the Smell Of It." She is packing into it what she has learned about perfume-making the world over, plus some of her own experiences pursuing her hobby. (End)

Akron Beacon Journal, October 20, 1968

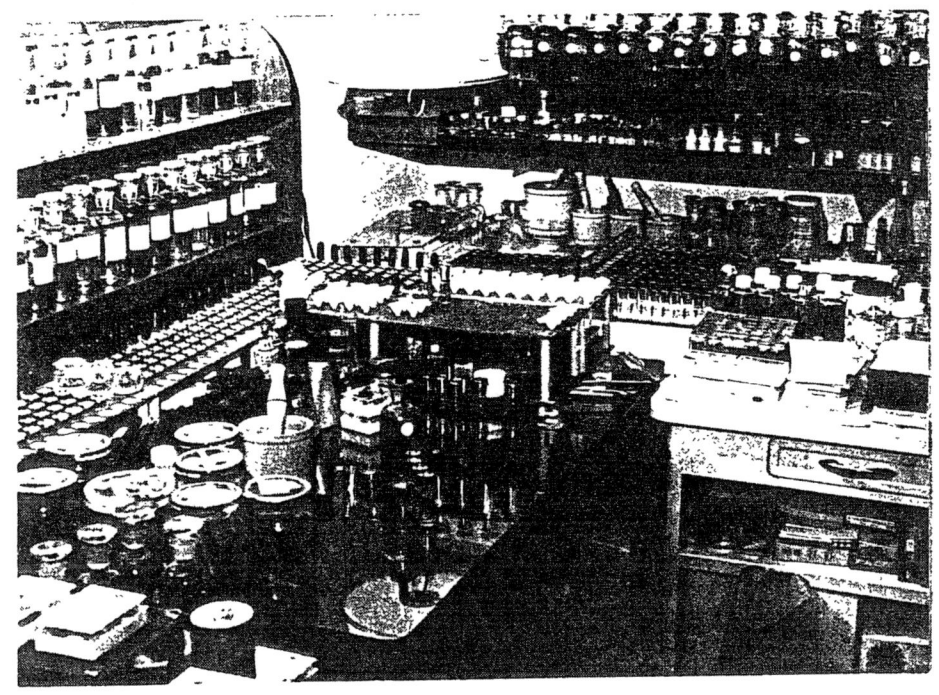

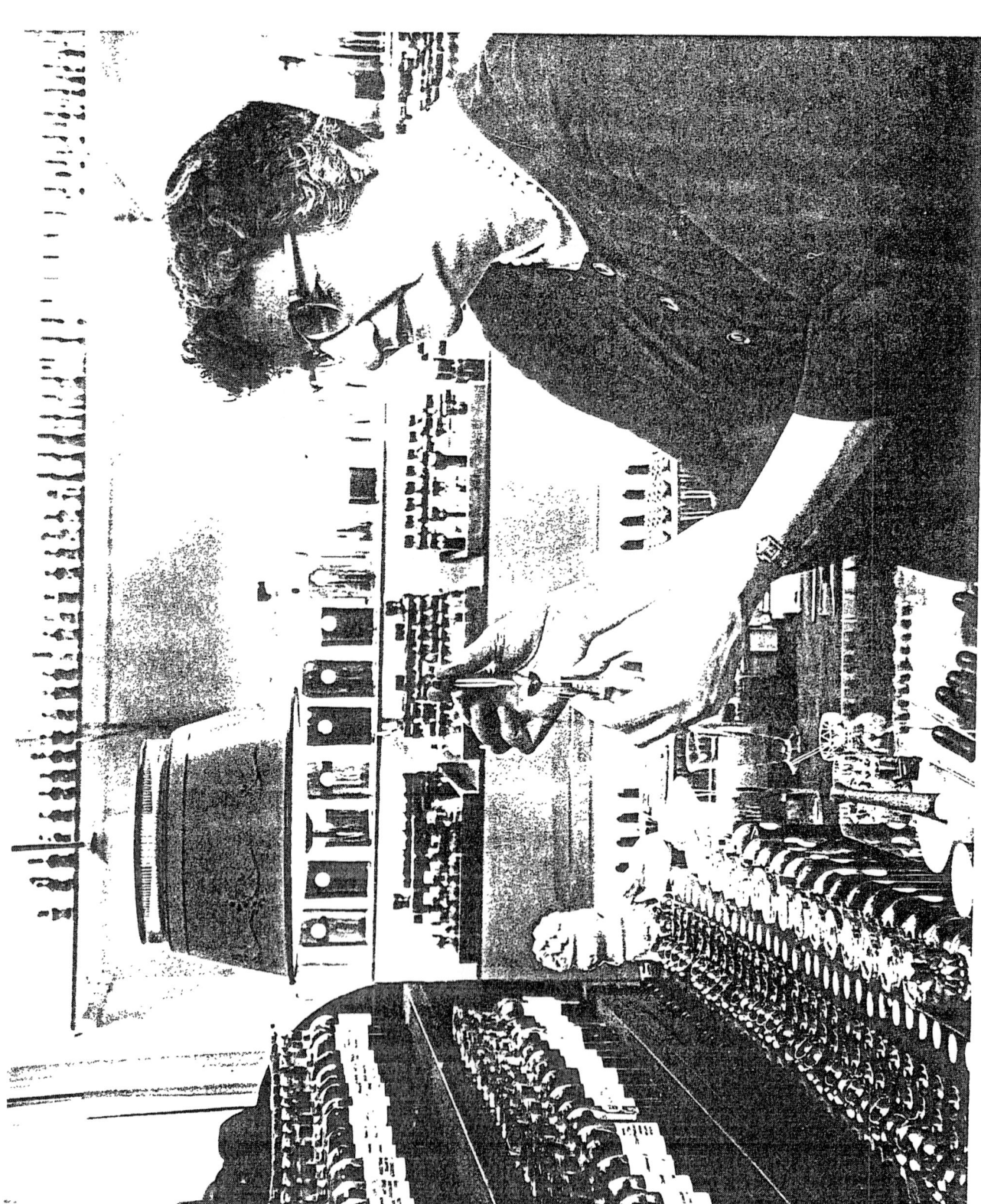

To:
> My Friend
>
> Fondly and fragrantly
>
> always
>
> Mary Lynne

I wish to dedicate this book to my husband, my mother and my dear friend, Dorothy Fuldheim.

Without their constant encouragement, this book would have remained forever a dream..never a reality.

Mary Lynne

Mary Lynne

Mary Lynne has received many honors during her forty years of research in "The World of Fragrance":

She lectured from coast to coast in the U.S., as a member of the "International Platform Association" on the following:

"The Romance of Perfume"
"Perfume of The Bible"
"The Art of Making Perfume Without Alcohol"
"The Art of Using Perfume to Advantage."

Her first appearance on radio was December 2, 1959, when she was the guest on "The Lee Adams Guest Show" in Cleveland, Ohio.

Thereafter, she appeared on radio with Dr. Blake Crider and Bill Gordon on the "Apartment 13" program. This was an Audience-Participation Program from 9:00 p.m. until midnight. The questions phoned in were a challenge and most exciting experience!

She then appeared on television station WEWS Channel 5 of Cleveland, with Dorothy Fuldheim and Bill Gordon on the "One O'Clock Club," which was another Audience-Participation Program. When the live studio audience were allowed to apply the different perfumes of Mary Lynne to their wrists, it immediately gained eager approval.

On August 12, 1968, Mary Lynne was the guest of internationally famous news commentator Dorothy Fuldheim of Cleveland Television Station WEWS for her "Noon Interview." At that time Mary Lynne made her "White Orchid" perfume on t.v. for the first time from the Vanilla Bean, which is the seed of the wild white orchid.

On August 18, 1971, Del Donahue, famous for his interviews on NBC Television Channel 3 of Cleveland, brought his film crew to Ravenna and recorded a tour of Mary Lynne's gardens, hom and lab during the process of making perfume.

On April 28, 1977, Neil Zurcher, who is famous for his "Travel Interviews" for WJKW Channel 8 of Cleveland, made a film of Mary Lynne's lab which aired on the 6 p.m. "News Program" the following evening.

On June 26, 1968, Mary Lynne was awarded the "Certificate of Merit" in London for "distinguished service as a perfumer and lecturer."

(cont.)

On April 12, 1969, she was selected as #455 of the "Two Thousand Women of Achievement for 1969" in the entire world. This place of honor has continued to be hers through the years.

She is also listed, under her full name of Mary Lynne Weenink, in:
"Who's Who of American Women, 5th ed. p. 1288,
"Who's Who in the Midwest, Volume 2 p. 1079,
"Dictionary of International Biography 5th ed. p. 566.

Preface

It is with pride and enthusiasm that I have the honor to introduce to the interested reader a book on the ancient art of perfume and fragrances, its integrated gentle growth and production, as it had been refined in ancient Babylon. Never before has there been a comprehensive compiling of facts based on personal work and experience.

But more so also on the gentle and very impressive personality of Dr. Mary Lynne. As an extremely busy pediatrician, she would find time to relax herself at night and gradually rediscovered and recreated the ancient art of perfume growth. From just a few flower petals, even dried, she created in her laboratory up to seven pints of genuine perfume of the respective flower, leaves or even grasses. Here most delicate and gentle warmth, combined with the trained mind of a scientist and practicing M.D., but also with the most knowledgeable close husband George, a trained biologist, helped bring about this genuine "Opus Minor," as the alchemists of the middle ages would have called it. It truly is an alchemistical grown essence of the flower, herb etc., used at its base.

As a serious practical student of this arcane "Royal Art," a fellow beneficial traveler of the "Art," Mr. Jack Glass from Ohio, introduced me to Mary Lynne in 1975. A close bond ensued, a bond of like minds, in pursuit of the ultimate Divine perfection. Mary Lynne did achieve this perfection in her "Opus Minor," her unique magical perfume creations. Even when she refused genuine help for her cancer ailment and it took over her body, she remained strong in her mind and a lady of the highest order.

Mary Lynne received many honors and had been a lecturer from coast to coast in the United States as a local and a national acclaimed personality. In 1969, she received in London the "Two Thousand Women Achievement Award of 1969" as one of the most outstanding women of 1969. Radio and television appearances in the late sixties and early seventies were considered a constant part of her life. Students of the "Royal Art" of alchemy both in the U.S.A. and Canada, as well as in New Zealand, Australia, Switzerland, Austria, England and Germany recognized at once her unique creations and achievements and paid her homage. I did have the honor of introducing her creations and method in most of these places, and still keep close contact to this day.

Mary Lynne has gone on in the meantime to the eternal Divine sphere, a difficult departure for most of us, but not out of reach.

In consequence, I am very proud and reassured that finally another fellow traveler of the "Royal Art," a good friend and lecturer of mine, Mr. Jack Glass of Youngstown, Ohio, as well as initially Mr. George Fenzke of Chicago, that both have taken the bold step to assist in publishing Dr. Mary Lynne in her life research work and practical results into the magnificent world of perfume and make it available to a wide audience.

May the Divine Guidance that guided Mary Lynne in her life work to rediscover these ancient secrets, also guide those that take interest in the miraculous world of perfume. To lift each one of us to these celestial spheres in our daily life and work, bring the mystical ancient, oriental world close to us, let us ourselves, open the Pandora's box, strike the magical lantern and express our three wishes which will be granted by God...without fail!

> Siegfried Hansch,
> Ottawa, Ontario
> June 17, 1994

"God give me the serenity to accept the things I cannot change,
the courage to change the things I can,
and wisdom to know the difference."

> Mary Lynne,
> (an old German proverb)

TABLE OF CONTENTS

1. The Beginning.......
2. "Mary Lynne" Was Born
3. Egyptian Perfumes Were Sacred
4. The Perfume of the Jews
5. Ancient Persians Loved Perfume
6. The Perfumes of Arabia
7. The Perfumes of Greece
8. Fragrance Among the Ancient Romans
9. Perfumes in the Life of Christ
10. Perfumes of China
11. The Fragrances of France
12. Perfumes of Spain, Italy and India
13. Perfumes of the Bible
14. Herbs
15. Other Sources of Fragrance
16. Fixatives
17. Methods of Extraction
18. Infinite Guidance
19. Incense
20. Fragrance of Famous People
21. The Attitudes of Perfume
22. The Art of Wearing Perfume to Advantage
23. Sniffing Can Be Fun!
24. Much Stranger Than Fiction!
25. The Ending.......

Chapter 1

The Beginning.......

Did you know God gave Moses the formula for making perfume? Do you know why? It is in the Bible. Can you find it?

Do you realize in 1967 $499 million was spent in the United States for womens' fragrances?[1]

They estimate that by 1970 this sales record will reach $689 million.[2]

Do you know some perfume today sells for $100.00 an ounce?[3] Do you know how much alcohol it contains? Just leave the bottle open for twenty-four hours and you will find out.

Do you know when--and why--perfume became diluted with alcohol or other bases?

Did you know Cleopatra made her world go 'round with perfume? You did? Well...do you know where she wore her perfume and why?

Have you read the story of perfume in Napoleon's life? Do you know his favorite scent?

Do you know why the ancients used perfume?

Do you know what makes scents?

---

[1] Beauty Fashion. June 1968
[2] Beauty Fashion. June 1968
[3] "Debutante de Versailles" by D'Andre per Beauty Fashion Magazine.

The answer to all these many questions await you in the World of Fragrance. However, to find the answers, and completely understand the value of perfume as well as its many uses, we must really start at the very beginning.

The first to discover perfume, it is believed by many, were the legendary inhabitants of Atlantis. Unfortunately, they left no written records of their era. Therefore, we must accept the hieroglyphics of the Ancient Egyptians as our first actual written record of perfume and its many uses.

The Ancient Egyptians used perfume for three distinct purposes: As offerings to their deities; for aesthetic purposes during their lives; and to embalm their dead.

The early Egyptians were extremely religious people. They celebrated at great length the birthdays of their many Gods and Goddesses. To favorably impress these deities, they constantly endeavored to make each celebration surpass the last.

Isis was, without doubt, their favorite Goddess. She was the Goddess of fertility and the harvest. The wealthy thought of her as the "Goddess of Plenty," while the poor worshipped her as the "Goddess of a Full Stomach."

Regardless of their personal feelings towards her, everyone from the highest Priest to the lowest peasant farmer joined in celebrating the birthday of Isis.

At one time, we are told in Ancient Egyptian history, the crops had been poor for several years, making it most difficult for the Egyptians in all walks of life. Therefore, everyone thought it wise to exert every effort to prepare a most unusual celebration for the birthday of Isis.

One of the Priests thought a burnt offering would be most appropriate for the occasion. So a contest was held during which the prize ox of all Egypt was selected. When the great day arrived, the ox was killed with great ceremony by the Priests. The carcass was placed upon a huge funeral pyre and the fires were lighted.

If any of you have ever smelled the horrible odor of burning flesh early in the morning, when most stomachs are not too stable, then you will realize the terrible experience of the Egyptians that day. Most of them became so nauseated they had to leave before the carcass had become half burned or the ceremonies were really under way.

This was most embarrassing to the Priests, as most of them became so ill that they, too, had to leave.

The entire affair was such an utter failure it left the people thoroughly frightened. All feared Isis wouldn't understand, probably resulting in a plague of locusts or some such disaster which would ruin their crops completely. However, Isis must have been a most understanding Goddess, for that year their crops were very good; even better than they had been for several years. Their good fortune, in view of the disastrous ceremonial, made everyone feel even more indebted to Isis. Therefore, the Priests immediately began to plan a more elaborate celebration for the next birthday of this beloved Goddess.

This time the Priests experimented endlessly until they made a remarkable discovery!

One of the Priests, while clearing out some of the underbrush which surrounded the Temple, discovered as this vegetation burned it gave off an extremely pleasant odor; in some instances, in fact, it was quite overpowering!

One vegetation of that day was the Eucalyptus tree. If you have lived in California, where this tree grows in profusion, you will recall when only a small handful of its buds are tossed into a fireplace and allowed to burn, within only a few moments the entire house is filled with a most delightful scent. The Eucalyptus tree was only one of the highly scented plants of those Ancient Egyptian days. The Priests were delighted with their discovery!

When the great day of the celebration neared, once again the prize ox was selected. But this time, before the carcass was placed upon the pyre, all the innards were removed and the Priests filled the huge cavity with the leaves, bark, roots and seeds, or seed pods, of the many plants, shrubs and trees with which they had been experimenting during the previous year.

When the people were all assembled, the fire was lighted. As it crackled and burned the hidden vegetation also burned, giving off a delightful fragrance!

Everyone was happy and gay! The celebrating, feasting, drinking and dancing continued on and on all that day, all that night and well into the next day. The Priests were in high favor!

However, they were not allowed to rest for long on their laurels, for the Ancient Egyptians were a people with many problems.

One of their problems was the Nile River.

In those days, the Egyptians depended on the Nile River for their entire water supply; for drinking, bathing and washing their clothing.

Each year there came a season when the weather became extremely hot and dry, at which time the waters of the Nile became low and quite smelly! During that season the Egyptians were able to maintain their health by drinking fruit juices, wines and milk, but if they were to bathe or wash their clothing, they had to use the water of the Nile regardless of its odor.

After the successful feast for Isis, they felt the time was right; therefore, they brought the problem before the Priests and asked them to search for a solution.

This they did.

Instinctively, the Priests seemed to feel the vegetation, which had brought them triumph through the celebration for Isis, would also enable them to maintain the respect of the people at this time.

During this period of great concentration and meditation, one young Priest decided to try an experiment. He gathered some of the same bark, leaves, roots and seed pods they had employed before. But, this time, he placed them all in a huge pot, poured some water over them and set it over an outside fire to boil.

He really had no idea why he had done this, it had never been done before; he simply felt guided to make the experiment.

The other Priests stood by and watched. All hoped and prayed for a miracle.

By nightfall the pot was still boiling away but no miracle had occurred. In disappointment, the Priests retired for the night.

The young Priest was the first to awaken the following morning. He hastened outside to peer into the pot. There he found a most surprising sight!

During the night, when the contents of the pot had ceased boiling and slowly cooled, a crust of hardened fats from the vegetation had formed on the top of the pot. Just as a covering of fat will form over a bowl of soup when it is allowed to cool.

They found this crust of fat could be gently lifted off the top. Even more exciting was the fact that this fat could be burned as it was or with the addition of a wick. As it burned, it gave forth a most pleasant fragrance!

The real miracle, the most joyous discovery, however, was the residue in the bottom of the pot!

They found when a bit of this highly concentrated water soluble residue was added to a bit of the water taken from the Nile, it completely covered the horrible stench of the river water, making it pleasant enough for bathing. This was surely the solution to their problem.

The Priests declared their finding, donned their most elaborate robes and, with great dignity, traveled up the Nile a short distance where, amid a great ceremony, they scattered the water soluble residue from the bottom of the pot into the river. By the time the water reached the people who had gathered below, they were able to wash their clothes and once again enjoy the water of the river for bathing.

This early experiment may not seem important to you, but in that one act the Ancient Egyptians founded two industries which have flourished throughout the ages and today are still considered leading industries.

The manufacturing of religious and scented candles or oils which give off a fragrance as they burn and the industry of masking.

Whenever we cook onions or cabbage and wish to remove the unpleasant odor from our homes, we use a household deodorant... a spray. In other words, we mask an unpleasant odor with a more pleasant odor. As we go about our homes, squirt, squirt, squirting from a little can of this cover-up fragrance, few of us realize this idea was originated in Ancient Egypt and perfected by the Egyptian Priests. This deodorant today is made from the water soluble residue which is left after the precious perfume oils have been extracted and sold to the perfumers. In the World of Fragrance, not a single scent is wasted.

As I have said, the Ancient Egyptians were a people with many problems. Another thing that bothered them was the fact they could not celebrate their funerals half as long as they wished. The climate was so hot they could maintain a body for only a few hours. This certainly was not their idea of a proper funeral.

Once again, they approached the Priests for a solution.

By this time, the Priests had carried on considerable experimentation with the precious vegetation they had found to be so helpful. They now had quantities of perfume oils (carefully skimmed off the top of numerous pots they had boiled) so, once again, they began to work on the problem at hand.

They found by soaking strips of linen cloth in the melted fats skimmed off the top of the pots, and tightly wrapping the bodies to exclude all air, the bodies could be kept for not only one or two days, but for a week or ten days during which they celebrated a most extensive funeral.

Thus the Priests solved another problem.

Only _we_ know how very successful they were in this endeavor. Whenever an Ancient Egyptian tomb is opened there is found at least one mummy. These bodies, or mummies as they are now called, are so perfectly embalmed they remain in a state of almost perfect preservation. Still, how many people, as they read of these findings, realize the embalming was done with perfume? And it really was!

One more problem plagued the Ancient Egyptians--that was their unhappy home life.

Once again, the now famous Priests were consulted. And

this time they were ready, willing and well able to solve this great problem.

In those days, when a young couple wished to marry, they announced the fact to the family of the bride-to-be and also to the family of the groom-to-be. At which time, both families set about preparing a super-sales talk in an attempt to persuade the couple to live with one family or the other. In those days, no young couple ever went to housekeeping by themselves. Well, even in those ancient times in-laws were not always compatible. Consequently, from time to time, there was considerable friction. The Priests, and men in high power, realized this could result in something extremely serious. They also realized Egypt could find strength only in unity, so all were most anxious to remedy this unhappy condition.

The Priests were ready! They had made the discovery that placing some of each of the scented plants in one pot was like putting all their eggs in one basket. Therefore, they had begun to experiment with only one or two of the various plants in the boiling pot at one time. As a result, they had come up with many different perfumes.

Their experimenting in this line paid off great dividends because they were now able to prepare a special perfume for each of the great houses of Egypt.

The Egyptians were instructed by the Priests to arise in the morning, bathe and dress but gather in a family group before eating. The Priests prepared a ceremony for them comparable to our Morning Worship. After this ceremony, a slave of the household quietly walked among the members of the family carrying a golden bowl filled with the perfume prepared for that particular household. The slave proceeded to anoint each member of the family with the perfume prepared by the Priests just for them. It was such a sacred rite, the Egyptians were awed with its dignity and felt they were so closely united they dared to cause no friction through the day. The following morning the ceremony was repeated, and peace reigned supreme!

Through their system of intermarriage, these families soon became small communities, the communities became small villages, and the villages soon became cities. Occasionally, one family would visit another family...which was almost certain to cause trouble.

But the Priests were on their toes every minute! They had even thought of a solution for this possibility of strife. When the outer guard beheld the approach of guests, he was to immediately announce the fact to the inner guard. Through this system, by the time the first guest reached the threshold, a slave would be waiting at the door with the golden bowl of perfume and each and every guest would be anointed with the perfume of that household. Quite naturally, the guests were deeply impressed, for it not only made them feel welcome, but assured them the host and hostess had taken them in as a member of their household for the duration of their visit.

Needless to say, peace and harmony prevailed throughout

the visit. We might even say the Ancient Egyptians lived happily ever after simply through their use of perfume.

Now let us look just a bit further: Who was the slave of the Ancient Egyptian household who did the anointing with the perfume carried in the golden bowl? The slave was a Jew.

During their years of slavery, the Jews had seen great and wonderful things happen for the Egyptians simply through the use of perfume.

They had seen the Gods smile down on them because the Egyptians offered up fragrance in their honor!

They had seen the bodies of the loved ones preserved by the embalming with the perfume.

They had watched the Egyptians enjoy great prosperity and live in constant peace and contentment, all through the use of perfume!

Therefore, they were inclined to believe the Priests may have been guided by the Gods and Goddesses in the preparation of their Holy Oils.

When the Jews were finally freed from bondage in Egypt, and allowed to return to their homeland, there was only one thing they asked to be allowed to take with them. That was just a tiny bit of the precious perfume they had seen do so much for the Egyptians.

They did not wish to use it for celebrations or funerals, or to embalm their dead. They only wished to offer a few drops in thanksgiving to their God for their new found freedom.

By this time, perfume was the most precious possession of the Egyptians. The Priests had taught them to believe the Gods directed the making of the perfumes. Consequently, the Priests had not passed up this terrific source of revenue. Their price for perfume was so high it had become even more costly than gold or silver or precious stones! For that reason, they would not allow the Jews to take one drop of their precious, and sacred, perfume out of the country.

We are told they went so far as to search each Jew to be certain no perfume was smuggled on their person.

The Jews went home sad and heavy-hearted. They felt surely their God would think them most ungrateful. But we are told the Jews are people of great faith, and they prayed without ceasing.

Finally, their prayers were answered.

When God spoke to Moses on the mountain, He gave Moses the formula for making perfume!

This formula can be found in the Bible...Exodus 30:34... There it tells exactly what plants to use and how much to use from each plant.

One night, after a particularly difficult day of trying to help solve the problems of others, I simply could not sleep. Just before dawn, in desperation, I took up the Bible from my bed-side table and began turning the pages hoping to find something which would restore my peace of mind.

Suddenly it amazed me to find in bold dark print at the

top of one page the word "PERFUME."

Intrigued, I eagerly read on to see what in the world would be found in the Bible pertaining to perfume.

In the thirtieth Chapter of Exodus, I found the actual formula. I had majored in chemistry, therefore, I immediately realized this formula was most unusual. There were listed seven ingredients. I knew seven was God's magic number, but wondered that a chemical formula should contain an uneven number of ingredients. We are taught today there is usually one catalyst to activate each synergist. In this formula, there was a catalyst to activate each synergist and then one extra catalyst to activate the whole. This is the secret of making perfume without alcohol!

My curiousity knew no bounds from that moment on. With the secret of the Ancients as my only equipment, I rushed forward, entering without fear the golden gates leading to a new and mysterious world...the World of Fragrance...where fact and fancy link hands for one purpose, and one purpose only: Simply to have fun.

Chapter 2

"Mary Lynne" Was Born

In the days that followed, my husband found in necessary to travel a great deal. He was busy teaching doctors and student doctors the wonders of Supplementary Nutrition in the treatment of Deficiency Diseases. His lectures took him from coast to coast, keeping him away from home for weeks, and often months, at a time.

Those were rugged days for both of us. Every moment of my life was filled to overflowing. From 8:00 in the morning until 11:00 or 11:30 at night was spent in the office or attending night school classes.

Throughout the war years, there were also nights of bandage making, canteen, etc., at the Red Cross. Bearing in mind I had no son to give for my country, I did all I could to help the sons of others.

All this necessary activity left me very little time for working in my precious World of Fragrance. However, we are told, "where there is a will, there is a way!" Consequently, I made every spare moment of my life work double time.

Each night I read myself to sleep. Often it was necessary to read a huge volume of the ancient history of a particular era from which I would emerge with only one or two paragraphs concerning perfume; but it was worth its weight in gold. Years later, all this proved to be priceless knowledge.

This was my preparation period!

The few moments daily I managed to escape from the blood and thunder of our war-torn world into the peaceful enchantment of my new found world gave me the courage and strength I needed to endure. I found my only relaxation and peace of mind in my blessed hobby.

Only God knows how eager I was from time to time to throw everything else overboard for the exciting future I always felt awaited me in this new World of Fragrance!

Then would come the thought of those who depended on me. The patients who, in gratitude, had sent me more patients; the doctors, with whom we had worked so many years, who had grown to depend on me for instantaneous research.

My conscience kept me steadfast. As a result, those few stolen hours spent in study of perfume meant as much to me as any lovers' rendezvous.

The wealth of valuable information gained from study only served to stimulate within me a burning desire to try my hand at making perfume. Words could never express the thrill of accomplishment which followed my first attempt.

Employing the method of the Ancient Egyptian Priests, and his boiling pot, I extracted various fragrances from both shrubs and spices.

The oil which floated free from the boiling pot of clove buds became the most sensational carnation perfume I have ever found!

My studies then took me to the field of perfume from floral extraction. This proved to be far more difficult. The actual blooming season in Ohio turned out to be extremely short. Hothouse blooms lacked the stamina needed for lasting fragrance.

It was then a famous cosmetic chemist and boyhood friend of my husband's came to my assistance. It was he who sent me my first samples of natural essences with which to experiment. He also furnished me with introductions to many importers of note.

Their great interest in my "perfume hobby" filled me with gratitude, humility and often embarrassment because it was not until many years later that my purchases from the importers were really of any significance to them financially.

Their constant interest and encouragement in my feeble little experiments never ceased to amaze me. It was not until many years later I learned this interest had stemmed from the fact that I was actually a "freak" in the trade; "one of a kind;" "something out of this world" and perhaps only "a flash in the pan," but someday might turn into a "gold mine" if they could only convince me to make perfume with alcohol, so they better not pass me by. I quote their exact words related to me long after I had "arrived."

Many times I have had to search for years trying to find an essence which gave me the true fresh fragrance of the flower actually held in my hand. Upon each occasion, the men in the field have moved heaven and earth as they joined in the search.

As they read this book, and I know they will, I want them to realize how thankful I am for their assistance. Without their help my hobby would not have been nearly as much fun, and my perfume would never have come into existance.

The only real friction I have ever had with my importers came about as a result of my refusal to add alcohol, chemicals or synthetics to my perfume. Some, in the early days, implored me to use my knowledge "wisely" because a fortune could be my reward.

But, come what may, I have constantly refused to work with alcohol. In the first place, it does not agree with my nose-- makes me sneeze! Also, I have felt dedicated to give the people of my day the same pure, unadulterated perfume God meant us to have from the very beginning.

As time went on, I learned much about the "oils" with which I worked. I found they had great enzyme value, as well as enzyme chain reactions when I finally learned how to hold them in suspended animation.

Another search led me to a man in New York who had just discovered how to make a precious little bottle with a spill-proof opening. Just the thing I needed to expel a single tiny drop of my finished perfume.

About this time I began to feel quite brave. For Christmas I boxed and gift-wrapped several bottles of my first blends and gave them as Christmas gifts to friends and patients who had everything else. As I had no labels, I hand decorated the bottles to make them appear more ornate. Well, then, believe it or not, my troubles really started!

Each one, who had received my perfume as a gift, came to me demanding that I sell them more perfume for their friends. I was overcome with panic; I had never dreamed what all of this could mean.

I was making perfume only an ounce at a time. Even this small amount took me weeks in the preparation. I had no labels, and I had no permit. What to do?

One more time, I appealed to my importer friends for guidance. They informed me I should first select a name, then have labels printed. Select blend names have tiny labels for the bottom of each bottle explaining which blend was contained in each bottle. It all sounded so simple...until I tried to select a name. My husband suggested I use my own name. "Mary" seemed too short, and our last name was far too long and difficult to remember.

After long and serious thought I decided to use my first name "Mary" and the last part of my middle name "Lynne." Thus, "Mary Lynne" was born!

My next step was to have foil labels of "Mary Lynne" printed for the front of my bottles. This was quite simply arranged. A dear friend did the art work. But the foil labels cost me over fifty dollars for a thousand. Later I learned it was wiser and much cheaper to have fifty-thousand made at one time. I learned slowly, and painfully, but I have never forgotten those lessons.

In selecting names for my blends such as "Persian Nights," the perfume made from the herbs of Persia (which gave such delight to the Persian Harems), I simply picked a name from out of my dreams, never thinking for one moment the name might be in use somewhere else in the world. Because everyone thought this was only a hobby, a plaything, no one thought to warn me.

A few years later, when I was sending my perfume near and far, I happened to mention to one of my new importers that I had received an offer from a well-known store who wished to sell my perfume for me. He very quietly said: "Be sure to have your name, and the name of your blends searched and copyrighted before you place a single bottle in a store. It would be extremely embarrassing to have to withdraw them because you were infringing on someone else's copyright."

This legal aspect was like a bomb! Why had I not thought of this? Guess I was having too much fun.

By this time, I had perfected several extremely interesting blends. For each I had selected a name which either recalled to my mind the story of its origin or the history of its purpose. Proper search revealed many of these names had been previously used, thus making it impossible for me to ever market some of my perfumes under their hobby names. Maturing from a hobby to a business caused my precious little perfumes to suffer many growing pains.

The very first search and application for copyright was

made for "Mary Lynne" and "Persian Nights." This required the assistance of an attorney and months of waiting. I was amazed at the cost of all of this! The urge to forget it all often entered my mind.

The growing demand for my perfumes by my friends and their friends was my greatest encouragement. I dared not disappoint them.

All this time, I had continued to give my perfume to others with my love. When my friends refused to take it for their friends, I was forced to set a price. This, I found, was very upsetting to me. How could I set a price--in dollars and cents--on something so near and dear to me as my fragrances? They insisted!

Finally, I sat down and figured the exact price per bottle as near as possible. Math had not been my major. So I simply selected $3.00 per dram as my price. On some bottles--depending on the ingredients involved in the blend--I made two or three cents per bottle. On others, I lost as much as five cents. My financial statement at the end of the first year of "selling" was a most pitiful sight, believe me.

It was then I found a solution which has given me great joy throughout the years. I would not sell my fragrances... I would share them with others at my cost with my love. From that moment on no woman in all the world has been happier.

Uncle Sam hasn't been happy, though. Almost every year he has noticed the amount of my total income and been shocked at the amount of total profit from same. His caretakers call me in for investigation. At first they bark at me, then, when I show them my books, explain that it makes me happy to do business even though I do not make money at if for I know I am making people happy every single day...the interview usually ends by the caretaker smiling, escorting me to the door and assuring me he will make a full and detailed report which will be in my favor. I then return to the lab, knowing for one more year I shall find peace and quiet.

Today my beloved little children, as the perfumes are to me, have reached their ultimate. In keeping with their progress, I have now dressed them in the sleekest of evening attire: stream-lined bottles designed with a most sophisticated air.

The foil labels have now given way to a golden banner bearing the name "Mary Lynne," which is gracefully screened from shoulder to shoulder.

How mature and promising they appear as they stand beneath their lovely golden caps, just waiting to bring happiness to others as they unfold the mysteries of their native lands. The emotional reassurance they were created to give to others pours forth with every drop.

Now with legal assurance that the world was my apple, I began over thirty years of the most fascinating research anyone could ever imagine. The stories I have unearthed have really been "out of this world."

Let us journey into the past and delve into the mysteries and magic of perfume.

Many countries, down through history, solved their problems with fragrance. Let us now go to Egypt!

Chapter 3

Egyptian Perfume Was Sacred

The Priests of Ancient Egypt were the first to make their perfume. They were known in history as the very first perfume blenders. This pursuit was considered a very mysterious and much esteemed art. It was their pleasure, and greatly to their advantage, to keep it so.

As we have already discussed, the Priests were the first to extract from vegetation the oils, resins and perfumes so popular in their day. They let it be known, as time went on, it was the Gods and Goddesses who really instructed them in this great and magic art. Because of this, the Priests could demand a high fee for their work. The Egyptians felt duty bound to also offer gifts to their deities, who so charitably came down and tutored the Priests, hoping to remain in their favor always. You can quite easily see, by this system, the Priests received double pay. Truly perfume was a God's send to them; they never had it so good!

The early Egyptians considered the possession and constant use of perfume as evidence of great wealth and social standing. Because of their value, perfumes were employed most lavishly during all regal functions and processions. It is recorded in history that one pharaoh, in honor of a great and glorious public spectacle, ordered twenty children to carry saffron, myrrh and other perfumes in beautiful golden basins, in addition to a whole train of camels each laden down with 300 pounds of frankincense, cassia, orris, cinnamon and spikenard. This represented such tremendous wealth even the the Gods must have been greatly impressed.

No Egyptian king could possibly be crowned without being first anointed with very costly perfume oils which were applied by the Priests, who pretended the ceremony was sponsored by the Gods.

As we have said, the mummies, or dead, were embalmed by means of perfume in those days, but that was not sufficient for the Egyptians. Each year the royal mummies were once again sprinkled with perfume.

When Howard Carter and his party opened the tomb of King Tutankamen, who ruled about 1350 B.C., there were found many receptacles for fragrant perfumes and ointments. The urns and vases were exquisitely executed in alabaster and solid gold, all containing quantities of aromatics which were still faintly fragrant after all those years!

The odor of these perfumes has been compared with coconut oil, broom and various resins. Chemical analysis supported the view that the ointments contained about 90% fat of animal character. The remaining 10% appeared to consist of resins and balsams.

From inscriptions on the tombs of the Ancient Egyptians, we are told all the people were required to perfume themselves at least once a week. Again, this was to help mask the body odors so common during that era.

The Ptolemy monarchs were extremely fond of perfumes and used great quantities of myrrh, saffron, cassia, frankincense, cinnamon and orris, which were their favorites.

Cleopatra, who was a Ptolemy, shared this deep love of perfume literally and figuratively, making her world go 'round with its delightful fragrance. By its magic enchantment, she won and held three most famous men!

Her most dramatic conquests, of course, were Julius Caesar and Mark Anthony.

If you will recall the story of Cleopatra's meeting with Julius Ceasar, you will remember it was most daring. When her brother, who was also her husband, turned upon her, she barely escaped with her life and remained in hiding for some little time. Her brother sent for Caesar, who was most anxious to add Egypt to his realm; Caesar eagerly accepted the invitation to visit Egypt. As they were plotting in the castle, they came upon the will of her father, the previous king, which definitely left the kingdom to Cleopatra and her brother. Julius Caesar realized it was she that he must win over, but how to find her?

At exactly the same moment, Cleopatra was trying to devise a way in which to meet Caesar without being killed by the guards posted by her brother.

At last she decided to approach the city from the sea long after dusk had fallen. Whether she reached harbor safely by bribing those about her, smuggled in by other boats, or quite secretly, it is impossible to say. It is sufficient to realize her undertaking was highly successful.

Apollodorus rowed the boat to a quay just below the royal palace. By that time, it was quite dark. It was then, in a little boat, that she prepared for her meeting with Caesar.

She applied fresh makeup and anointed herself thoroughly with the perfumes and unguents which were her pride and joy. Then she instructed her Greek companion to tie her up as though she were a bale of merchandise. This he did and bound a rope securely around her so that it was impossible for anybody to suspect how precious was the treasure he held in his arms. It was most fortunate that Apollodorus was strong and powerful while Cleopatra was light and slim, for he flung his burden over his shoulder and bore it into the royal palace without attracting too much attention.

The spies of Pothnius were far more alert and dangerous than the Roman sentries. The ruse turned out to be highly successful.

Apollodorus, in accordance with his request, was conducted directly to Caesar and, dropping his load in front of him, he loosened the bonds, unrolled the covering and, to the amusement of all present, the Queen of Egypt sprang to her feet in all her glory.

It was impossible for Caesar to remain unmoved by this unparalleled feat of daring. She had planned every detail to impress his mind and fire his imagination. The sudden and unexpected sight of such a fascinating young woman, the appearance of the Queen, whom he was anxious to meet because of political reasons, as well as the tremendous risk she had taken to reach him filled Caesar with admiration.

The shocking surprise, her great beauty, her charming manner, her delightful voice and the overpowering perfume she had applied so lavishly completely took Caesar off his feet!

Julius Caesar had the reputation of being a devil with the ladies. But he found more than his match in Cleopatra. She had spent years preparing herself for this encounter.

She had even ventured so far as to have a "still-room" made, wherin she spent days at a time distilling unguents to conserve her beauty. There she made the perfumes to gain her the favors of men, like Caesar; she made beauty, charm and intrigue her business.

Some historians write much concerning her beauty, others say she was not really beautiful but extremely attractive. All speak of her charm and her delightfully musical voice as "tinkling bells" and "soft and sweet as honey."

During her experiments, she had discovered that by adding colors to her precious perfumes, which were in oil form, crude paints were produced. The practice of applying "face paints" reached an all time high during her reign. The very highest degree of cosmetic art was attained in concentrating on the eyes. The underside of the eye was painted green while the lid, lashes and eyebrows were painted in a very heavy black.

This page was intentionally left blank.

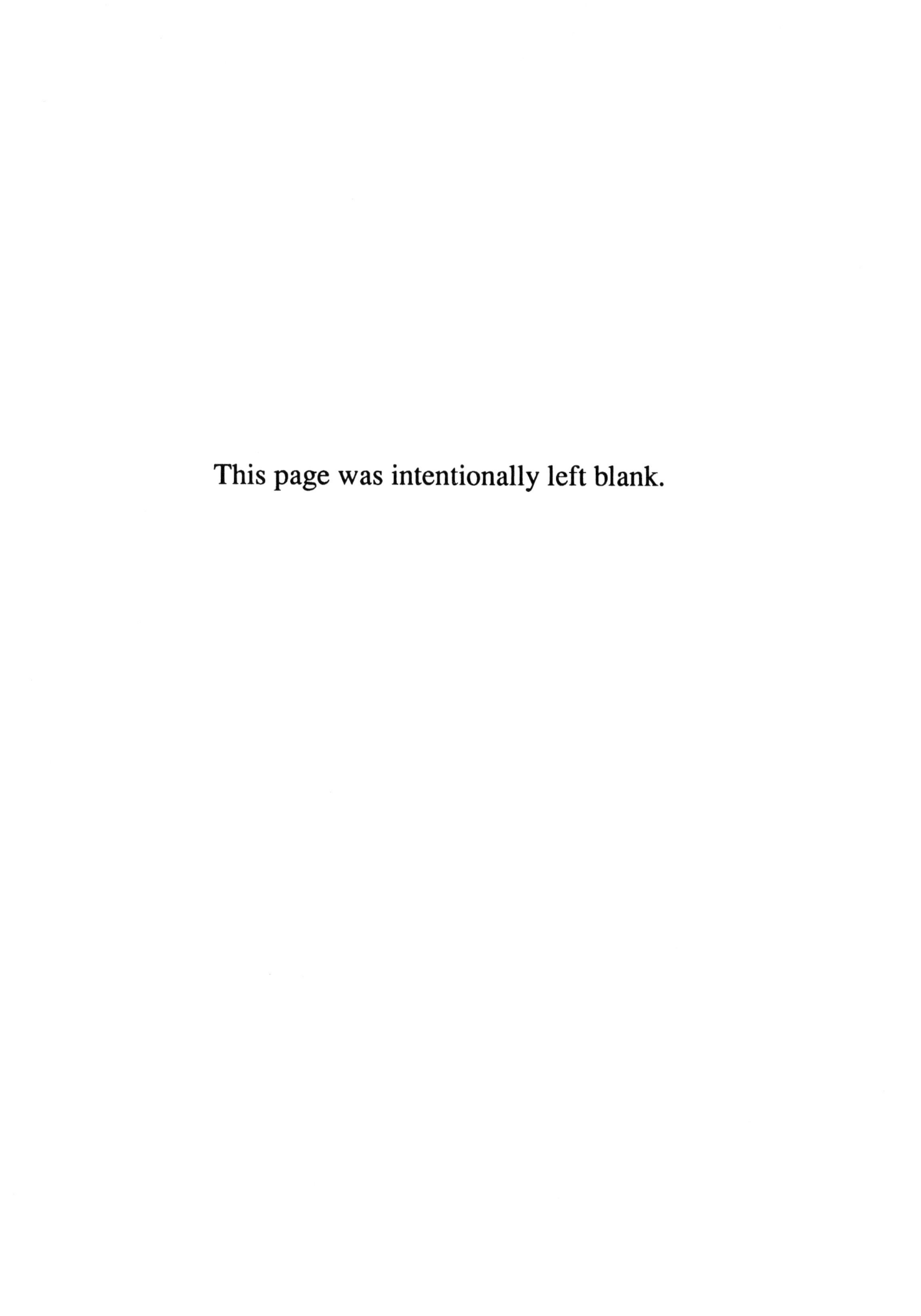

Today it is the fashion to highlight the eyes with heavy make up when the lady has a nose which is too large. Some historians, who had managed to remain sober in spite of her charm, tell us Cleopatra's nose was not beautiful and was inclined to be larger than it really should have been. How clever she was to conceal this one defect in her great beauty! Again we are reassured "there is nothing new under the sun" even in make up.

Caesar was helpless when with her and miserable when away from her. To know her love, her charm and to lie drunk from the effect of her perfume was all he asked of life at that time. However, he remained under the spell of his new found love too long for his own good. When he was forced to return to his own country and leave her in Egypt, he could never forget her.

Upon Caesar's return to Rome, he seemed extremely displeased whenever anyone approached him wearing perfume. Historians have gone so far as to say "Caesar hated perfume!" I believe he hated it, especially in the later years of his life, because it reminded him of Cleopatra.

At this time, Rome was just beginning to use perfume lavishly--even the men wore it upon almost every occasion.

We are told whenever one of Caesar's soldiers approached him wearing perfume he would fly in a rage. He severely punished many soldiers for this bit of bad judgement. Once, we are told, before his personal army he declared, "I would far rather my soldiers reek of garlic than to go about smelling like women!"

Later on, much was written regarding the romance of Cleopatra and Anthony. There, again, she used perfume to win his love and hold it.

You will recall, when Anthony sent for Cleopatra to come to him, she decided to go by barge or boat. She was clever enough to realize it was not wise to rush to his side, so she kept him waiting just a bit. Anthony grew impatient because of his tense expectation. Then, finally, came the day!

Asia's all-powerful ruler, Anthony, was sitting in judgement in the market place at Tarsus. He was surrounded by nobles and dignitaries and multitudes of other people. It seems the market place was teeming with humanity. Then Anthony saw how the passive, listening crowd suddenly became animated. All were in motion, as people whispered something hurriedly to one another and in a second the market place was empty.

Everyone was running towards the river, chanting as they ran, "Venus is coming! Venus is coming in festal procession to visit Bacchus!"

From the banks of the Cydnus there rose the scent of precious perfumes. Cleopatra, who was really the Venus of the Nile, had come to Asia Minor to celebrate fresh triumphs.

One of the most beautiful passages of literature, pertaining to perfume, was written by Shakespeare as he described Cleopatra and her barge as she went to meet Anthony: "I will tell you./ The barge she sat in, like a burnish'd throne,/ Burn'd on the water. The poop was beaten gold;/ Purple the sails, and so perfumed that/ The winds were lovesick with them..."

Could anything be more romantic?

It is interesting to note that it was the Egyptians who first discovered that citrus fruit peel could be pressed until odorous oils oozed from it! This crude method of extraction is displayed in their ancient hieroglyphics.

The very oldest of all surviving perfume trademarks is the Ancient Egyptian name "Kuphi." Kuphi was the most sacred of all religious fragrances. Each night it was offered by the Priests in the temples to please the Gods.

Cleopatra, who lived in the first century before Christ, believed Kuphi should not only be confined to the Gods. She felt women should be allowed to wear this perfume daily.

History does not tell us who won this argument, but we are told she wore much perfume of "her choice."

The story I have always found most amusing is the one which tells us when Cleopatra found a man to her liking, she placed her perfume upon her lips and then kissed him; thereafter he never left her side.

Remember Julius Caesar, Mark Anthony and now poor Mr. Burton!

Chapter 4

The Perfume Of The Jews

The second people listed in the long history of perfume are the Jews. This luxury did not appear to have come into common or general use among them until long after their return from Egypt.

During their long captivity in Egypt, they first became exposed to the many refinements of that country. They silently watched the Egyptians enjoy their many luxuries. Secretly, no doubt, they had visions of having all those lovely things for their own. Many believe this association changed the Jews completely, from a very simple people to an extremely polished and most industrious nation.

Long before that time, we are told, the Jews had carried on some experiments and had discovered the aromatic properties found in the gums and resins of shrubs grown in their native land. However, they did not really appreciate their true worth until, while held in captivity, they saw the Egyptian Priests perform miracles through the use of the sacred perfumes they prepared in the secret chambers of the temples.

As we have said, upon their return to their homeland, the Jews constantly prayed for a perfume with which to thank their God for their new found freedom. It was then that God spoke to Moses and told him many things.

God instructed Moses exactly how to make the precious perfume (Exodus 30:34) which was to be used only in the temples.

He emphatically stated the perfume made from this sacred formula must be kept holy for the Lord. In Exodus 30:37 and 38, we read: "It shall be unto thee holy for the Lord... Whosoever shall make like unto that, to smell thereto, shall ever be cut off from his people."

For that reason, I have never tried to reproduce this precious formula. I have no desire to set a precedent because, you see, I like it here!

The art of speaking in parables, that is, by indirect and concealed expression, was a fashion with the learned Jews of ancient times. Also, we find almost everything had its symbol. Therefore, from the direct translation of the ancient Hebrew, we learn the words of God passed on to Moses held great and hidden meaning for the Jews.

Let us pause a moment to reflect upon a translation I have found of these words.

<u>Principal Spices</u>-(and there were three): To represent the composition of the Holy Spirit.

<u>Of pure myrrh</u>- Symbol of wisdom--knowing what to do.

<u>500 shekels</u>- The amount of myrrh was equal to the amounts of cinnamon and calamus combined so God, in our anointing, gives us wisdom equal to our understanding and knowledge combined.

<u>Of sweet cinnamon</u>-Symbol of the understanding which we have of all the knowledge we get from God.

<u>Of sweet calamus-</u> Symbol of the knowledge which we receive from God's Word.

<u>250 shekels-</u> The amounts of cinnamon and calamus were equal, showing that God by his Holy Spirit will give us an understanding of whatever knowledge we obtain from His Word.

<u>Of cassia-</u> Symbol of deputyship, described in other passages as "workmanship" and as "counsel-and-might." God gives no knowledge except for a purpose; He establishes our faith by giving the understanding thereof, then gives us, and finally blesses us, with the advice and ability (counsel and might) neccessary for good workmanship in the accomplishment of His purposes.

Pure frankincense-Frankincense is the symbol of praise, heart adoration.

Throughout that particular line of study, I could but wonder what happened to the Jews.

God thought enough of them, and them alone, to give them the formula for making perfume. Indeed, they were his Chosen people.

The overpowering significance of the priceless emotional value of this formula was given to them simply in answer to their prayers.

The Jews remained depressed, still reeling from their many years of slavery to the Egyptians. The fragrances of plants mentioned in the formula which God gave to Moses has an uplifting effect upon those who inhale the perfume. It offers "mind expansion," new horizons, new hopes and new dreams! Why did God select those seven plants when there were so many thousand plants available? I believe it was because of the superior enzyme chain reaction rate of those seven plants.

Yet, somewhere along the line of history, the Jews simply discarded this heavenly gift, completely overlooking the secret it concealed which could made them the most powerful, the most famous, and the most wealthy of all races. Why?

Was it the simplicity of the formula which eluded them?

Often we find, among the Jews especially, the complex offers the greater challenge.

But not so with the Persians.

Chapter 5

Ancient Persians Loved Perfume

The Ancient Persians adored perfume! To use it as offerings to their Gods, or as a means of embalming their dead, did not interest them in the least. They loved perfume for the feeling of sheer joy and ecstasy it gave them as they wore it and inhaled its delightful fragrance.

One of the most interesting stories of perfume in Persia was one which related to the Crusaders.

We are told the early Crusaders of Briton devoted their lives to sailing around the world. They gathered up beautiful and different things of interest, culture and refinement with which to improve their own country.

Once upon a time, while on such a voyage, the Crusaders finally reached Persia. By this time they were a long way from home and extremely lonely. Soon one of the better salesmen of the group managed to talk their way into one of the harems for a night of relaxation and fun. They had a wonderful time!

As they emerged from the harem the next morning, they were comparing notes (as I find men are very apt to do). They all marveled at the splendor of the previous night's entertainment. During their conversation they were puzzled as they recalled, rough and tough Crusaders that they were, that they should so completely relax with such a large group of perfectly strange women! In comparison, they remembered these

Persian women were inclined to be rather short, dark and decidedly on the heavy side. They could not speak a word of their native tongue and, above all, they were not too clean. Water there was mighty scarce, at times, and not to be wasted in bathing.

On the other hand, they remembered their ladies at home. They were taller, fair as the golden sunlight, slim as willow twigs and clean beyond all comparison, for they swam every day and loved the water.

In spite of this great contrast, each Crusader had to admit he had never had a better time in all his life. After considerable discussion, the question in their minds still remained: Why was this possible?

It was then the youngest Crusader of all spoke up. "You know, I am the youngest of the lot. I had never done anything like this before. It was so new to me, it was frightening. So, when the rest of you rushed into the harem, I stood back and was the last one to enter. When the curtains closed behind me, I was so scared, I took a big, deep breath to give me courage; from then on, nothing mattered!"

The Crusaders were quiet for a moment, all thinking mighty fast trying to recall every detail of their entrance into the harem the night before. Then they all began to talk at once!

Each one remembered there was a smokiness in the air and a very definite fragrance about the place none of them had ever encountered before but, in their eagerness, they had disregarded it at the time.

They were all so stirred with curiousity that they knew there would be no sleep for them until this question had been settled once and for all.

The only solution was for all of them to return to the harem on a thorough tour of investigation--and another night of fun!

This time, upon entering, each Crusader looked about, suspiciously sniffed the air, then each man took a very deep breath filling his lungs with the sweet scented fragrance. Sure enough, it worked like magic! It had happened again. Within seconds each man was completely relaxed, oblivious of all problems or worries and fired with the desire to have fun.

On the second night, however, before they completely fell apart with exuberance, they were clever enough to ascertain what had caused this terrific feeling of relaxation. The girls of the harem were delighted to explain, as best they could, and displayed to them the wonderous potion.

They brought forth beautiful bowls made of gold and alabaster set with precious stones. All were filled with amber colored oils which continuously wafted into the air a heavenly fragrance. They pointed out tiny silver lamps filled with glowing coals hanging here and there throughout the harem. They explained that each morning they sprinkled a few drops of this precious perfume on the glowing coals because, as it burned, it filled the air with happiness. They continued by explaining each night they anointed their breasts with this perfume because it gave them happy dreams all night long.

The Crusaders were enchanted! How wonderful it would be to take this new discovery back to their homeland.

Needless to say, all the Crusaders left the harem the following morning with a tiny golden or alabaster bowl filled with the precious oils a much wiser and happier man.

Sailing homeward, each realized much had been accomplished for culture during this visit to Persia. Each Crusader had acquired a precious treasure, had learned the use of the new word "perfume," and his alone were the memories he would cherish forever.

When the weary Crusaders reached Briton, they eagerly presented the perfume to their ladies who exclaimed with delight at its heady fragrance.

"It is lovely, but what is it?" they asked.

"It is precious "perfume" from Persia," the Crusaders replied.

"But what do we do with it?" the ladies persisted.

In answer, the Crusaders explained the perfume could be burned in tiny lamps filled with glowing coals, which seemed a terrible waste of anything so precious. Besides, they didn't have any little silver lamps. Therefore, they instructed the ladies to apply the perfume to their breasts at night exactly as the women of Persia had.

Well, the ladies of Briton could hardly wait until nightfall!

Finally, it was time to retire. In the household of each Crusader the supreme experiment was about to take place. The perfume was applied. Each lady found its fragrance to be most pleasing and quite exciting. But when those gentle ladies beheld the effect of the perfume on their menfolk, they were positively amazed and highly suspicious.

They demanded to know just where the men had secured those perfumes!

The Crusaders were so dumbfounded by this onslaught of jealousy they all quite innocently told the truth. They simply said they had secured the perfume in the harems of Persia. Nothing more.

This explanation seemed sufficient at the time to quiet the ladies. However, I like to think a few days later, perhaps at a Ladie's Aid meeting, the women began to discuss the one subject uppermost in their minds: perfume. It certainly was the conversation piece of the day.

Suddenly, during the discussion, they were all surprised to realize not a single one of them knew what a harem really was. So, to a woman, each one rushed home and the Crusaders were given the third degree. By that time, though, the clever devils had gotten their heads together and had prepared a very simple story.

"Life is very different in Persia than here in Briton,"

they said, "if this were Persia, we would call a harem a little perfume shoppe!"

Just to prove how resourceful these men were: history tells us that not a single lady ever learned the real truth about a harem because in those days women never traveled. Each one lived and died believing a harem was only a delightful little perfume shoppe!

The more research I did pertaining to the perfumes of ancient Persia, the more fired I became with the desire to reproduce not only the fragrances of their time, but also to capture the emotional reactions they experienced as a result of their perfumes.

My tireless efforts were well rewarded, for I finally learned there were fifty-nine precious oils combined in the perfumes of the harems. Patiently, I searched for these for many years. Upon receipt of the last one, I eagerly set about blending them together in the proper combination.

As soon as they had aged for thirty days, I knew I had been rewarded a thousand fold for all my effort. I had captured not only a heavenly fragrance, but all the emotional deliverance known to the ancient Persians within the privacy of their harems.

You see, those clever Persians had used a combination of fifty-nine particular scents which they had extracted from herbs to relax all complexes, do away with all inhibitions and give those who inhaled the fragrance one desire--to have fun!

This one voyage made by the Crusaders to Persia did wonderful things for the history of their homeland.

It gave them a new word: perfume. It gave them a new product: perfume. But, best of all, it gave them a new, more luxurious way of living to which they have clung down through all the ages.

This perfume I have christened "Persian Nights." I have shared it at my cost with my friends, with their friends, and with those who have heard my lectures. It seems it can be worn with a most interesting effect. Some enjoy wearing it during the day...others reserve its magic just for evening wear...it all depends on just what you expect of it. Let it be sufficient to say, I have a file filled with letters from all over the country telling me the miracles this perfume has performed.

When entertaining at home, I have found great fun in placing one drop of my "Persian Nights" perfume on one electric bulb in each room. As soon as the bulb becomes warm, the entire room is filled with the same relaxing fragrance known and prized by those of the Persian harems...more fun!

Chapter 6

The Perfumes of Arabia

In the centuries that followed, the Arabs seem to have delved much deeper into the serious side of perfumery than any other race.

Of course, the Ancient Arabians needed perfume for an entirely different reason. Their problem was B.O. (body odor). Most of the men of their day were sheperds who spent their entire lives tending their flocks. Many of them never had a bath in all their lives, since water was far from plentiful.

At length their body odor became too much for even them to bear. About this time, one of the sheperds, while killing a young goat for food discovered a tiny gland filled with a semi-solid oily secretion. He must have been in the mood to experiment, for he rubbed this secretion on his skin and immediately realized he had made a terrific discovery!

The application not only completely covered the almost overpowering body odor, but it also softened his skin which was weatherbeaten from the hot sun and strong winds.

This good news spread quickly from one sheperd to another until finally all were enjoying this new and inexpensive type of perfume.

There was only one person who simply could not stand it! He was Avicenna (980-1037), the Arabian doctor and chemist who cared for these particular sheperds when they became ill. To him this new found odor was unbearable. The body odor had been bad--but this was worse!

One night, after an exceptionally difficult day during which he had cared for many of the ailing sheperds, the doctor felt he simply must walk for a time in his little garden until he could free his nostrils of those most distasteful odors. As dusk was falling, he lingered beside his beloved rose bushes breathing deeply of their heavenly fragrance. Suddenly, to his mind leapt the thought, "Why couldn't the sheperds have made an oil of this delightful fragrance with which to rub their bodies?"

Avicenna was an interesting person. The fact that he owned two precious roses tells us he was a gentle and most generous man. For it was necessary for him to share his daily ration of precious water with his rose bushes or they would never have grown or bloomed at all!

As he inhaled their fragrance, he continued to play with the idea that surely there must be a way to capture this fragrance. His mind then wandered to the little still with which he distilled his herbs into medicine for his patients.

Distillation in those days was very crude--nothing compared to our modern methods--but it did accomplish one thing: it changed solids into liquids.

He was fired with an idea! He set up his little still, ran out and plucked all the petals from his precious roses and hurriedly ran them through the little still. When all his rose petals were exhausted, the doctor had only a few drops of highly concentrated but delightfully fragrant oil. He was so excited, he ran from one friend to another asking them for their rose petals to carry on his experiment. After all the rose petals available were exhausted, however, he had less than half a teaspoonful of the fragrant oil.

It was most discouraging. He realized Arabia could never grow sufficient roses to make this oil, which he called "Attar of Roses," in any quantity. So, with a heavy heart, he began to dismantle his little still. Piece by piece he washed it in a tiny bowl of precious water. To his delight, he found this water soon took on the lovely fragrance of his roses as they had grown in his garden.

Even though the doctor was never able to reap any reward for his discovery, he made careful note of each step of the experiment and years later chemists and perfumers made use of his notes when making Otto of Roses...and Rose Water.

The perfume extracted from the rose was, probably, the first of all the floral scents. Down through the ages the fragrance of the rose has remained the favorite of all. Much has been written concerning its influence throughout history; it has been mentioned more than any other flower in literature. The rose goes hand in hand with romance.

Today, the lovliest "Attar of Rose" perfume is made in Bulgaria.

In the early hours of the morning, long before dawn, the

peasant women of Bulgaria, dressed in their bright colors, go singing into the rose fields of the "Valley of Roses," to gather the fragile petals from thousands of roses.

The petals must be picked, loaded into carts and taken to the distilleries before sun-up in order to capture the elusive and magic fragrance of a rose still alive and growing on its bush. When even slightly touched by the sun, the same petals give off a wilted, depressing odor...such as we notice upon entering a funeral parlor. This attitude must be avoided or the entire purpose of the rose fragrance would be defeated. Therefore, great haste is made to procure the "Attar of Roses" we all adore.

Just imagine! It takes about four thousand pounds of Bulgarian rose petals to produce one pound of rose oil. This oil often sells for $2,000.00 per pound to the perfumer; best of all, it is worth every "scent" of it!

Even today the water, with which the equipment is washed after each distillation, is known as "Rose Water" and is used in hand lotion.

Each time I work with the fragrance of the rose, I think of Dr. Avicenna and his desire to help the sheperds of his day. Many times I find myself wondering if he can possibly know how much his simple little experiment has helped us all.

Chapter 7

The Perfumes of Greece

It was in the fourth century before the Christian era that the conquests of Alexander the Great first brought perfume to Greece.

In Greek mythology, perfumes played a very important part. They believed the Elysian Fields were not only fashioned of various types of delightful perfume, but they pictured in the midst of these fields a glowing city of solid gold.

The gates opening into the city were supposed to have been made of fragrant cinnamon. Around the walls surrounding the city of gold there flowed a river of precious perfume. All who dared to approach were enveloped by the intoxicating fragrances rising from its waters.

The Greeks firmly believed when Phaon, the Lesbian pilot, carried Venus of Cypries his reward was a present of divine perfume which transformed him from an ugly looking fellow to a man of godlike beauty!

They were also certain that Circe held fast to Ulysses by means of a perfume which was filled with magic.

Venus was supposed to have been the very first user of aromatics, and man's knowledge of them was attributed to an indiscretion of one of her nymphs who served as her handmaiden. Paris thus conveyed to Helen of Troy the secret. Her fabulous beauty was supposed to have resulted from her constant use of this secret and enchanting perfume.

Through all the ages the sense of smell has been considered the most ethereal of all the senses. In fact, some of the Ancient Greek philosophers have gone so far as to place the soul in the olfactory nerve!

When we recall the history of ancient Greece, we first think of marble...then we think of perfume because culture and perfume have always walked hand in hand down through the ages.

Magallus was an old perfumer made famous in Greek history by his discovery of a rare fragrance which was known as "megallium." The lasting aroma of this substance is most delightful!

Peron, of Athens, was a well known perfumer of his day. He was even mentioned in a poem thusly:

> I left the man in Peron's shop just now
> Dealing for ointment; when he has agreed,
> He'll bring you cinnamon and spikenard essence...

Plutarch wrote:

> The soul of a man in love is full of
> Perfumes and sweet odors.

In the days of true Greek luxury, perfume held a value almost equal to that of food. In the description of one famous banquet, we are told each guest was first required to engage in a ceremony called "the purification of hands" done with perfume.

Each guest was presented with a small gold or alabaster container of sweet odors.

As they ate, a fine spray of costly perfume was showered down upon them. There were also beautiful white doves, with wings heavily saturated with rare perfumes constantly flying above the heads of the guests, causing the perfume to fall gently as they fluttered their wings.

It was for this particular reason that at this banquet something never before heard was brought forth: Myrrhine, which was a combination of myrrh, honey, sweet smelling flowers and wine...which they drank!

One of the favorite dramatists of that era gives us a delightful description of the toilet of a Greek man-about-town of that day:

> He really bathes
> In a large gilded tub, and steeps his feet
> And legs in rich Egyptian urguents;
> His jaws and breasts he rubs with thick palm oil
> And both his arms with extract sweet of mint;
> His eyebrows and his hair with marjoram,
> His knees and neck with essence of ground thyme.

This gives us an idea of how completely addicted the Greeks became to perfume. Unfortuneately, they carried the use of scents to extremes. They had scents for each portion of the body. Scents were designed and used to clear the befuddled minds. Perfumes were designed to cure every known illness of the day and, of course, scents which were guaranteed to win love at any cost.

The Greeks became such addicts to perfume that Solon executed a law prohibiting the sale of perfumes. This could have been disastrous had not the edict been so unpopular it was openly and glaringly violated constantly. It is written even Diogenes visited the perfume bootlegger's speak-easies where he lavishly perfumed his feet. He declared it was most wasteful to use such precious perfume on one's head, for the scent rose in the air and only the birds flying above could enjoy it. By using it upon the feet, however, his entire body was bathed in the delightful odors as they rose heavenward.

The Greek physician Hypocrates was the first to prescribe perfumes for his patients, particularly for those suffering from nervous disorders. Using this as a basis, one Greek poet wrote, "The best recipe for health is/ To apply sweet scents on the brain," and, "The rose distills a healing balm/ The beating pulse of pain to calm."

In many countries perfumes have been employed for medicinal use down through the centuries. Musk has always been regarded as a curative for almost any ill in China, especially headaches, nervous disorders and even snake bites. In the healing art, the rose fragrance exceeded all others.

Homer frequently mentioned perfumes in both his "Iliad" and "Odyssey." In the former he describes the toilet of Juno in these words, "Here first she bathes, and/ Round her body pours/ Soft oils of fragrance."

Theophrastus was, no doubt, the earliest Greek writer on the subject of perfumery. He was born in 370 B.C. His outstanding work was on botany. From him we learn particular oils of his time were obtained from flowers. This is the first appearance of floral oils in the written history of perfume.

The Greeks enjoyed perfume to the fullest. Even the dead of those days were provided with perfumes in their passing. A flask of their favorite scent was always placed in the coffin. For those too poor to afford this luxury, a picture was painted on the casket of a perfume container.

The Greeks of ancient times raised perfume to their high level by accepting it as one of the arts of culture and dignity.

Chapter 8

Fragrance Among The Ancient Romans

During their early history, the mighty Romans displayed very little interest in perfumes. This disinterest was furthered by an edict passed about 188 B.C. forbidding the sale of any exotics. It was not until their migration into south Italy, then occupied by the Greeks, that they acquired a more intimate knowledge of the aesthetic side of life. Then the Romans followed the Greek culture in all its many arts and elegancies. Perfume was no exception.

In the hey-day of the Roman Empire, anything which was expensive and difficult to obtain was the vogue.

As time passed, the Romans even outdid their Hellenic neighbors in the lavish use of scents at feasts and such functions. It was their custom to have great showers of perfumes fall from the ceilings of the buildings upon the guests at such occasions.

Nero became Emperor of Rome in 54 A.D., and then perfume oil assumed a most important role at his court for he simply adored it! It is written that Nero was so very fond of perfume he used more in a single funeral procession than could be produced in Arabia in ten whole years.

In this era nothing could possibly overshadow the fame of the Roman bath. In keeping with the exotic tendencies of that day, among the plutocrats especially, bathing was a sensuous rite. Among the upper classes a great deal of the social life of the day was centered around the marble bathing pools of their large estates. There they received their friends and guests. It is written, "they gave themselves over to lucurious dalliance, held indescribable orgies and interspersed these diversions with frequent dips into pools of water exhaling every variety of seductive perfume." After these dips, we are told, hosts and guests alike lay on marble slabs and were massaged by slaves with scented oils and balms.

Some very wealthy Romans went so far as to use a different scent or essence for each part of the body.

Their perfumes were "rhodium"--a great favorite--which was made by steeping rose petals in huge vats of wine; "melinum" made from quince blossoms; "metopium" from bitter almonds and "narcissinum" from the narcissus blossom. Other fragrances of that time were more complicated. Some were quite elaborate. The King of the Parthions had prepared for him one perfume which contained twenty-seven different fragrances blended into one.

Men wore perfume profusely--almost to extreme--and all garments were perfumed. Horses and dogs were rubbed with fragrance in an ointment form. Military flags were perfumed. All public fountains, as well as the small fountains in the homes and on the estates, were perfumed each day. Even during the most bloody encounters in their famous ampitheater a soft mist of saffron fell like dew from the tent-like roof.

Once upon a time, when Catullus invited his friend, Fabullus, to dinner, he chided him for being stingy by telling him if he wished to eat he would have to bring his own food but added, "On the other hand, you shall have from me love's very essence, or what is sweeter or more delicious than love, if sweeter there be; for I will give you some perfume which the Venuses and Loves gave to my lady; and when you snuff its fragrance, you will pray the gods to make you, Fabullus, nothing but nose."

During this era, perfume was so much in demand entire streets were lined with perfume shops. The perfumers of that period were actually worshipped and considered indispensable to all mankind. These men were extremely talented chemists, clever artists and true magicians who plied their magic ruthlessly at times to further their own gains.

The Romans were the first to discover the secret of making glass. Therefore, they were the very first to place their perfumes in glass bottles. These bottles, however, had no caps, tops or stoppers until many years later, when alcohol was first added to perfume.

Licinus and Julius Ceasar forbade the sale of perfumes (but even they were unable to curb its use.) Julius Ceasar, evidently, never forgot (or forgave) the fact that it was Cleopatra's perfume which held him to her side when better judgement told he should have returned to his own country--where great trouble brewed.

From this chapter in history, we note the ancient Romans luxuriated in perfume because it was the most expensive luxury of their time; expensive because it was very difficult to obtain.

To them, perfume was only one more addition to their licentious way of living. It stimulated them! They accepted its soothing qualities as merely a sedative which allowed them to forget their consciences.

How amazing it is to realize at that very moment perfume was being used in a most Godly manner!

Let us see how beautifully it was brought into the life of Christ...the son of God.

Chapter 9

Perfume In The Life Of Christ

We are told, on the night which stands out above all others in the history of the world, King Herod sent forth three wise men to find the newborn prophet.

The wise men followed the star in the East which led them to their destination, the Christ Child born in a manger. They fell to their knees before Mary and worshipped her son.

They also presented her with the gifts which had been sent by King Herod--an offering of honor to Him, the Son of God.

What were these precious gifts? They were the most costly gifts to be had...gold, frankincense and myrrh.

According to the Venerable Bede, who was an English writer of the seventh century, the second wise man was named Kaspar.

He is described as a beardless youth, with light hair and a very ruddy complexion. His gift to the Christ Child was frankincense, a fragrant gum which, when placed on hot coals, produced an aromatic vapor to rise.

Such a gift was in keeping with Oriental custom and served to emphasize the kingly esteem in which the Infant was held.

Myrrh was a symbol of wisdom. This was included among the gifts not only because of its great value but also to signify a gift of great wisdom throughout His reign as the Prince of Peace.

As we touch upon perfume in the life of Christ, I wish to tell you the story of Mary Magdalena, who was also known as Miriam of Migdal.

It seems, she was a woman of two very different lives. When she was in her home in the Kidron Valley, or the Lower City, she was the woman of the world. The most popular prostitute of her day. She drew to her all the rich and cultured youth of Jerusalem.

In her home in the Valley, she entertained her wealthy admirers by dancing for them--more beautifully than any other dancer of her day. Her body was breathtaking in its beauty, draped only with the thinnest of silks and the long flaming veil of her hair which, as she danced, gave glimpses of her nakedness here and there.

Men came from far and near only to be completely intoxicated with her charms. She became a very wealthy woman exceptionally deft in her trade.

High upon the Mount of Olives, she had a second home in which lived her sister, Martha, and her brother, Eliezer. It was there, near this home, that Mary Magdalena had her beloved garden.

Planted in this garden was a collection of rare herbs, weeds, spices, plants and blossoms which she had collected from all the four corners of the world.

Whenever she heard of a rare flower or plant in some faraway place, she immediately sent for it and had it planted and cared for in her garden. Nothing was more important to her than that she possess every new and delicate odor.

Fragrances delighted her so highly, she made it her business to learn all the intricate means of extraction and compounding of perfumes.

In those days, every woman of aristocracy had her individual perfumes and ointments prepared for her, but Mary Magdalena surpassed them all! Quite by accident, she had come upon the secret of making perfumes which were absolutely irresistable. She bathed in these precious oils several times daily and also soaked her sheer veils, with which she danced, in the heavenly fragrance.

Her garden attracted scholars of all nations. Frequently visitors from Rome, Greece and Egypt, who happened to pass through Jerusalem, would request permission to visit the garden and study her many rare plants. Visitors, however, were seldom admitted, for she desired to reserve this place for herself. Often, when she became weary of her gay life in the Valley, she would retreat to her garden of spices to spend whole days there in meditation.

As she meditated and worked with her perfumes, she was constantly haunted by problems of eternity, realizing more and more that she should have a purpose in life.

Often, when troubled with these thoughts, she would disappear for weeks at a time during which she would fast, repent and purify herself. Her next step would be to seek the council of a Rabbi, and ask how she might wash away her sins.

It was during one of these repentent periods that she learned the new and wonderful Rabbi (Christ) was visiting the house of Simon the Pharisee.

The servants were unable to restrain her as she rushed into the rooms of Simon where the learned men were reclining at a banquet asking many questions of Christ.

In her excitement, she was a thing of great beauty! Her body was draped with many-colored delicate stuffs of Zidon (such as our chiffon.) Her hair, like leaping flames, was not concealed, as the custom of the day, but fell over her neck and bosom. There hung on her body by a golden chain an alabaster vial containing the most delicate of perfumes. The scent quickly spread throughout the entire house.

All present at the scene sat in utter amazement!

Suddenly she stretched forth her hands, asking, "Who among ye is that wonderful man who hath called to all and sundry, 'Come unto me ye that are heavy-laden and I will give you rest?'"

She carefully perceived each man. At length her eyes came to rest on Christ. She drew near Him, and fell at his knees weeping bitterly.

Her tears fell on Christ's feet and washed them. When she saw her tears had wet his feet, she took her hair and dried them. Then she kissed His feet with her lips.

She then removed the alabaster vial, which hung at her throat, and poured the precious perfume it contained on the feet of Christ and anointed them.

Within her heart, beneath the silken raiment, Mary Magdalena was filled with repentance. Christ realized at heart she was not really a sinful woman but had been the victim of her own experiments with perfume. Therefore, He forgave her and requested that she "go and sin no more!"

Never again, we are told, did she dance before men who were thirsty for love, nor did she ever again wear her perfume to draw men to her. She continued to grow the plants and extracted their fragrance, but during the rest of her life she sold the perfumes to women to help keep their husbands at home--and gave the money to the poor.

Throughout the New Testament it is often mentioned that Christ was anointed with perfume or he anointed others. This was the custom of that day because the perfume used for anointing was considered a very Godly thing.

At the close of His life on earth, we are told when they removed the body of Christ from the cross, Mary and her helpers wrapped the body in cloths which had been soaked in frankincense and myrrh.

Therefore, Christ's life on this earth began and ended amid the fragrance of these two lovely perfumes.

Chapter 10

Perfumes Of China

There is no exact date given in history to tell us just when China first became interested in perfume. They have merely recorded, "Perfume has been used in the Orient since the beginning of time." And, after all, who knows when time for them really began?

However, we are told, from the beginning, the Orientals' interest in perfume was most unusual. First of all they required something far different from perfume than any other race of people. They desired a perfume which would stimulate them and excite them sexually.

Working towards this goal, it was the Monks of Tibet who were the first to discover "Musk." This animal scent gave new "life," new "desire" when inhaled.

It was soon realized perfume containing this new ingredient, called musk, not only caused humans to desire the perfume for their own, to have and to hold, but also the one who was wearing the fragrance!

In other words, when the Chinese discovered musk, their ambitions were realized and their mission accomplished.

Confucius tells us, "Incense perfumes bad smells, and candles illumine men's hearts." Therefore, incense is used in all Chinese temples as well as in their homes, representing a most important part of their daily worship.

The Chinese not only inhaled their fragrance but they also drank it!

We are told that from the millet a most delicious and fragrant spirit is made and consumed as a part of their ceremonies. Perfumed joss-sticks, as well as tinsel paper, are used in China to such a great extent the preparation of same is a thriving industry.

For many years, China had but one perfume which they called "heang." It was rich in musk because they believed musk could cure any disease as well as create desire, thus keeping one both well and happy.

The writings of Peo-po-tse, a noted Chinese physician, tell us he often prescribed musk as a preventative and also as a cure for snake bites. In the records he has left behind for our research, we find he has called musk "shay hiang." Could their heang and his shay hiang have been the same?

People who traveled in the mountains of China were strongly advised to insert as much musk underneath the nail of each big toe as could possibly be packed there before starting their journey. This was believed to protect them from snakes, serpents and reptiles in general. It seems snakes just cannot stand the odor of musk.

Strange as it may seem, the Chinese were not interested in making new and different perfumes. They used, and greatly enjoyed, sandalwood, patchouli and musk. Beyond that, they were just not curious.

Among the fragrant flowers they used for the temples and in their homes were the mo-lu-hwa. This resembles the jasmine in appearance and scent. Just one tiny flower is sufficient to fill an entire room with a most overpowering fragrance.

The kwei-hwa, the chu-hwa and lien-hwa are all extremely beautiful and scented flowers they also enjoy. Still, there is no record of the Chinese ever making perfume from any of these lovely blossoms.

It has always seemed to me, the Chinese knew what they desired, they found it, they were thankful and their search was ended. Musk filled all their needs, so why exert themselves to search further?

How clever--how wise the Chinese!

Chapter 11

The Fragrances Of France

History tells us perfume was first introduced into France by the Crusaders. From then on, especially after the twelfth century, the history of perfume became the history of France.

With the exception of only one or two, all the Kings and Queens, even the uncrowned Queens of France, were patrons and patronesses of perfume. We are told in 1190 Philip Augustus granted a charter making a Frenchman the very first licensed perfumer. This opened the door for the perfumers to become recognized men of a trade. The art of perfumery, which previously had been considered more magic than art, now became a dignified profession. Naturally, this did much to elevate the standing in society of both perfume and the perfumers.

Charles V (1364-1380) had a great and almost uncontrollable passion for delectable odors and fragrances. Because of his great desire to obtain only the very best, he became the first manufacturer of perfume in France. This new interest was of such importance to him he proceeded to plant scores of acres of lavender, roses, lilies and violets in his gardens and spent long hours in his laboratories every day experimenting with new blends and combinations previously unknown.

The Ladies of his Court felt free to really let themselves go as far as perfume was concerned! They eagerly sought the fragrance of musk and ambergris, for well they knew how both attracted the men. All cosmetics of the day were perfumed to high heavens! Even perfumed soaps were peddled through the streets, the vendors loudly calling forth their wares. It is interesting to know that during this era the French imported their scented soaps from Naples.

During this period, tiny sachet bags, called "coussines" were made of lovely satin filled with lavender and other soft fragrances, which were hung among the clothing and placed in wooden chests with the linens.

For the first time, books on perfume began to appear. We are told one of the very first was "Les Secrets de Maistre Alexys le Piedmontois," which went so far as to give a recipe for a magical water guaranteed to make women beautiful forever, "Take a young raven from the nest, feed it/ On hard eggs for forty days, kill it, and/ Distill it with myrtle leaves, chalk and/ Almond oil." Amazing, isn't it?

Simon Barbe even went so far as to open a school where he taught the art of making perfume--the first of its kind. His sign tells us he taught the methods of extracting the scents of all flowers, of preparing all kinds of perfume, the secret of purifying tobacco and perfuming it with any kind of scent. He accepted for this course of training those who entertained the nobility, religious persons, bath-house keepers and hair-dressers. Needless to say, he was an extremely busy person and very popular.

In 1693, Simon Barbe published a terrific book on perfumes. It was called "Le Parfumeur Francois." As this book became better known, it greatly increased his perfume sales. Many had never heard of perfume or appreciated its great value and importance until they saw all the advantages in black and white.

The most interesting thing about this book, as far as I am concerned, was the preface. Here Barbe declared he believed he had contributed greatly to the glory of God through his perfumes. How? Because he had taught many of the clergy, "les personnes religieuses," to make perfume which could "cast spells." Again, we have the good and the bad.

Nevertheless, this book brought Barbe such royal returns that he soon wrote another which he called, "Le Parfumeur Royal." It was most unfortunate that just about the time this book began to be known King Louis XIV decided to banish perfume.

This brings us to the many struggles perfume went through for a mere existence in the history of France.

When Louis XIV (1643-1715) was King of France, he was affectionately known as "The Sweet-smelling Monarch" which, of course, was a terrific break for perfume in that day and age.

It was most unfortunate, however, that quite suddenly King Louis decided the terrific headaches, from which he had been suffering, must have been caused by his never-ending indulgence in the potent perfumes of that time. Therefore, perfume was banished from the Court! That is, all perfume except Orrange Blossom. This particular scent was deemed most essential in the promotion of romance. After all, they could not risk a sudden decrease in population, so the gentle fragrance of the orange blossom was allowed to continue with its work.

No one dared approach the King wearing the slightest trace of perfume. Because the King no longer loved fragrance, he felt the entire world should hate it so.

This made it extremely difficult for the perfumers of France at that particular time. They barely existed for a while and the country, as a whole, lost much of the happy carefree attitude for which it had become so famous.

Then Louis XV (1715-1774) became King of France.

He was very fond of perfume! It made him happy to wear it, so felt his subjects should not be deprived of the joy it brought them as they wore it. Consequently, once again perfume came back into its own.

During the reign of Louis XV, Madame de Pompadour set great standards of taste and style. All the arts flourished once more. Perfume was used so profusely during this era that the court of Louis XV was known both far and near as "la cour parfumee." What a happy time for both the perfumer and his fragrances.

Versailles became a place of entertainment and enchantment. The elegance, in every respect, far surpassed anything the French had ever known. We are told those of the court were requested to use a different kind of perfume each day. Madame de Pompadour, in her own household at Choisy alone, spent about 500,000 livres--which would be about $100,000.00 in our money--in one year on perfume. What a joy to have been her perfumer!

The recipes for many of her favorite perfumes, so famous in those days, have been handed down from father to son and are still in the possession of a well-known perfumer even today. It is said the original recipe for one was handed down to this firm by the heirs of Manon Foissy, who was chambermaid to the Marchioness.

Madame Du Barry was another shining light in the history of French perfume. It is declared that the great magician, Cagliostro, prepared her secret of beauty which was a most delicately perfumed lotion. She guarded her perfume secrets most carefully. In fact, she had her own "still-room woman" who distilled fragrant waters for her own private use. It seems money was no object as long as the perfume accomplished its purpose.

Those certainly must have been gay days for France! There were strolling perfumers in the streets who dressed themselves in scarlet coats with every elaborate gilt facings who were considered the elite of all vendors. Some even rode about the streets in most elegant chariots and sold their perfume to the accompaniment of a brass band. They sold the very best of perfumes, elixirs, opiates and cologne. Their business flourished, the French loved this bit of excitement, and visitors were thrilled with the very sight of it all.

Then Louis XVI became King, and, believe it or not, his physician convinced him that all this should be banished! He thought the gaity, the strolling perfumers and their brass bands were most undignified.

This physician, evidently, held great persuasive power over the King because the law was made according to his wishes. So once again, perfume had to run for cover. But this time the people did not approve.

As a result, perfume was boot-legged and sold at a very high premium because it was illegal and difficult to obtain. The perfumers did not suffer this time, you see, they had learned to be crafty through bitter experience.

The Renaissance had brought a marked revival of all the arts, so once again perfumery had flourished. Italy became the center of the industry at that time. For that reason Catherine de Medici learned at a very early age the tremendous value of perfume.

She bathed in scented waters, even as a young child. When she was old enough to carry a kerchief it was always scented with her favorite perfume.

Therefore, it was a most natural thing when she was taken to France to become the bride of Henry, who later became Henry II, for her to take not only her perfume with her but also her own private perfumer.

History merely tells us this famous perfumer was "Rene the Florentine," whose shop was on the Pont au Change.

Rene was a man of many talents. He was not only the very best of all perfumers, but was also one of the outstanding chemists of his day. He has the distinction of being the only person in all of history who ever blended perfume and poisons. This he did for his beloved Catherine.

During Catherine's first years in France as the Duchess, the wife of Prince Henry, she quietly observed and endeavored to follow the customs of her newly adopted country. She noted most of the Court Ladies frequently visited astrologers, whose business also included the sale of charms, perfumes and all sorts of ointments and love potions. In other words, the French Ladies of the Court relied on the French magicians to guide them.

In this respect, Catherine differed. She was afraid to trust the French magicians with her secrets and her fears because she knew the French did not care for her and spoke of her as the "Italian woman" behind her back. At the same time, she felt Rene was not quite able to meet the needs she felt in order to win the love of her husband with perfumes and potions.

Catherine's problem was Diane of Poitiers, mistress of the Chateau d'Anet, who had been Henry's mistress since he was about fourteen years of age! It is said, Diane had no less a personage for her beauty doctor than the famous alchemist

Paracelsus--and what an excellent job he did!

It was then that Catherine, becoming desperate, consulted the famous brothers, Cosmo and Lorenzo Ruggieri, who were Italian astrologers, crystal gazers, chemists and perfumers.

The brothers did much for her. Their greatest favor was when they kept her from poisoning Diane! However, try as they might, no one seemed able to compound a perfume of sufficient power to break up the love affair between Henry and Diane... it lasted until his death. It was then that Catherine became Queen.

Because of Catherine's frustrated love-life, she had concentrated all her efforts in becoming powerful in her own right. To gain her objective, from time to time, she never hesitated to use perfume--or poison.

Some historians tell us it was Rene who first designed perfumed gloves. I am inclined, however, to believe Rene was the first person to introduce perfumed gloves to the French Court. He had lived in Italy where making articles, especially gloves, from perfumed leather was a real art.

Nevertheless, Rene was certainly the very first one to make poisoned-perfumed gloves. Of this we can be certain.

Whenever Catherine was anxious to promote a favorite project in the French Court, the French resented it very much. It was her desire to rule France with a high and mighty attitude, but the French were of the impression she was a foreigner and should know her place. Therefore, someone would always have the courage to speak out against her project.

Whenever this would happen, Catherine, who was able to take any amount of insults without losing her temper, would simply leave the Court and retire to her own apartments. Then she would send for Rene.

The next day the one who had dared to speak up against Catherine would receive a beautiful package delivered by the messenger of the Queen. The surprise would always be complete. As the recipient accepted the package, they would always say, "See, after thinking the whole thing over, Catherine has seen the light, she realizes I was right and this is her way of saying she is sorry."

Which proves the French did not really know Catherine!

Upon opening the package, there would be a most beautiful pair of pastel leather gloves scented with the favorite perfume of Queen Catherine. The first thought would be, "Does Catherine know my size? Will they fit?"

Breathlessly, the recipient of the gift would rush his hand into the glove trying it on for size.

Very cleverly hidden within the seam of one of the fingers would be a needle which Rene had soaked in deadly poison for days. As soon as the needle ripped open the flesh, the deadly poison started its fatal mission. Within two hours the recipient of the lovely gloves would lie dead on the floor!

Needless to say, it took only a few such incidences to prove to the French that Catherine meant to be boss and should not be interferred with no matter what she decided to promote!

As time went on, it is said, Rene became even more clever with his perfumed gloves. He was most successful in blending poisons and perfume with the leather, and in this way could help Catherine in a definitely more sublte way. The person had only to wear the gloves for a few hours and they would die--the poison was taken in through the pores of the skin.

This gives you the horrible side of the association between Catherine and Rene. They both had their good points and later on did much to benefit France.

Definitely on the credit side of Catherine de Medici's dark account we have the beautiful institution of Grasse. It was Catherine, we are told, who, with Rene, did much research and investigation in that area. Finally they realized the area surrounding Grasse was ideal for the growing of flowers which would bloom the entire year round. So it was she who sent Sieur Toubarelli there to found the very first perfume laboratory--or commercial perfumery.

Shortly after this, the little town of Grasse sprung into fame as the seat of the natural flower perfume industry and it has held that distinction through all the years. Even today it reigns supreme.

Nowhere in all the world is the climate and soil better

suited for the cultivation of flowers in infinite profusion than along the beautiful French Riveria. That stretch of coast bordering the Mediterranean is one of the rarest garden spots in all the world. Here millions and millions of flowers are grown for their precious perfume each year.

It is true, Cannes is famous for its rose, acacia, jasmine and above all for its rare and delightful tuberose. But the most famous of all is Grasse, for there flowers are grown and immediately the perfume is extracted the whole year round.

There are well over fifty distilleries of perfume in the area of Grasse. Almost 100,000 acres of that region are devoted only to the growing of flowers.

The inhabitants of Grasse are noted to be the happiest people in the world. From the peasants who gather the flowers in the fields at early morning to the very learned artists of perfume who from them make the ethereal perfume essences there is constantly a song in the heart of each!

We are told Grasse was destined to be the "Garden of the World" because it is naturally protected. Its soil and climate are ideal for the culture of fragrant flowers. These glorious flowers, which give us such heavenly fragrances, grow as easily there as weeds grow in other climates. It is amazing to learn, though, that only flowers can grow there--it is not at all suited for general farming.

One famous writer states, "In Grasse men do what other men

the world over only dream of doing! Lucky people! Their entire life is spent in a Paradise of Flowers."

One account tells us that Grasse has over fifty perfume distilleries which net a yearly average of many millions of francs. Statistics tell us there are over 60,000 acres given up to the cultivation of flowers which produce some 2,200,000 pounds of roses and 3,300,000 pounds of orange-flowers. In addition, 2,200,000 pounds of jasmine and 220,000 are cultivated there each year. 25,000 pounds of roses are needed to produce one litre of essence which is sold for about 2,500 francs. We must realize these statistics may vary from year to year, but it does give us an idea of the percentage of perfume produced per acre.

Every month of the year there are flowers harvested in Grasse. From January to March, violets, jonquils, narcissus and mimosa are commonly harvested.

From April to June, roses, especially the rose centifolia, orange blossoms, migonette, pinks and Golden Broom are harvested.

Lavender, jasmine and tubrose are harvested in July and August.

Mint, geranium and aspic reach their peak during the months of August, September and October.

From September to December Cassia turns the entire countryside into a delightful sea of yellow and clouds of most betwitching scent.

Returning to the story of Catherine, Henry II was her

first love, but perfume was certainly her second. During her days as Queen of France it was good taste to perfume rooms, even public places, for special occasions. Catherine even went so far as to have the waters of the public fountains perfumed for festive events.

Following many of my lectures, guests have often come to me to say they had visited Grasse and found its great beauty to be exactly as I described. They have often added that they were privélleged to visit some of the distilleries and perfumeries of that area.

Still not a single one of them realized the equipment actually used in the distillation of the most expensive oils is the same equipment placed there during the reign of Catherine de Medici. The French guard this precious equipment as they would guard their lives. It is not only valuable to them but is a sacred trust placed in their care. It has made Grasse famous, given each of them a happy and contented life, but even more it has placed a song within the hearts of each and every one who has been given a part in the industry of perfume.

Rene will long be remembered for his perfumed gloves, "Gloves as sweet as damask roses,/ Masks for faces and for noses,/ Bugle bracelet, necklace of amber/ Perfume for a Lady's chamber."

Catherine de Medici and Rene will never be forgotten in France for the perfumeries of Grasse are a living tribute to their great love of perfume.

Chapter 12

The Perfumes Of Spain, Italy And India

Spain

It was the early people of Spain who discovered the cell construction of animal hide was identical to the skin structure of human beings.

With this thought in mind, they did considerable research pertaining to the embalming methods of the ancient Egyptians and soon began experiments by using the same methods to cure animal hides.

They soon discovered they could cut off the hair, or peel off the layer of hide bearing the hair, soak the hide in the perfume oils and the result gave them a most durable new material, which they called leather.

The most delightful thing was that this new leather which they had discovered was perfumed as it was preserved!

A new fad was born. Before long it was the fashion to wear gloves, jackets, jerkins and even long coats made of this precious (and very expensive) scented leather.

Spain was soon known far and wide for this new industry. The scented gloves found their way to England and France. The most popular of all gloves were those made of mouse skin, perfumed with a most delicate fragrance.

Thus Spain made a famous contribution to the history of perfume.

Italy

Not to be outdone with the discovery of scented leather in Spain, Italy immediately started to work on the same line of research.

It is said by many historians that the Italians even surpassed the art of the Spanish in making scented leather goods. They even bound books in scented leather. They also made bookmarks of scented leather. But they went even further in also dying the scented leather in most delicate pastel colors.

Of course, Rene, the personal perfumer of Catherine de Medici, who was Italian, went still further by making his gloves in pastel colors, beautifully scented, and also poisoned when the need presented itself. His gloves, needless to say, did not become the fashionable thing to wear. They were only to be used in case of an emergency. Wearing them only once was sufficient!

India

India has always been a country famous for its trees, shrubs, and many plants from which were derived many perfumes of great and lasting fame.

The unforgettable fragrance of their sandalwood used so extensively when building their Hindu temples is one example, as are the famous gates of Somnath which are still living reminders of their art.

History tells us the earliest caravans were loaded with spices, perfume and perfume yielding materials en route from India to Arabia. It was over 3,000 years ago that the Indian "nard" found its way into Egypt.

In the temples of India great fires were lit, composed of fragrant woods. The flames were kept ablaze with the the addition of consecrated ointments. All around the sacred fires was scattered a scented herb called "Kusa." Even today, this herb is used in several of the heavier, deeper perfumes.

It is written that India's paradise, or Swarga, is a haven for lost or unhappy souls. Within this paradise is found a perfume which is called "Calalata." It is supposed to enchant the senses and make any wish come true.

According to Hindu mythology there is not one but five heavens. A particular, and wonderful, God presides over each heaven. All five are filled with perfume and heavenly scented flowers.

In the Brahmin heaven is said to be found a most beautiful flower of blue--the "blue campac." The delightful fragrance of this one flower seems to be sufficient to fill all needs--therein.

High on the summits of the Himalayas, in the heaven of Indra, there is still another delightful flower to be found, the "Camalata." Its color is rose, and its fragrance most enchanting to all.

The stories written about the fragrance of India's many perfumes are many, and often quite confusing, especially the stories pertaining to their heavens and the promises awaiting there. It is a well known fact, however, that perfumes and fresh flowers even to this day play a most important part in the every day life of India.

Otto of Roses, the true fragrance extracted from the petals, is a very sacred perfume there. It is used in wedding ceremonies, given as gifts of distinction, and has remained for centuries their favorite scent.

At the Krishna festival, perfumes are sprinkled over everyone in the belief that the scent will wash away all impurities and sins for the past year.

A very similar custom is still followed in Latin American countries during the annual carnival or Mardi Gras celebration.

Here, again, we note the emotional reaction to perfume. It's a release from sin and depression, replaced by gaiety.

Chapter 13

Perfumes Of The Bible

When we are children, it is a bit difficult to picture clearly in our minds the true graphic description of the Holy Lands. Often the land which gave us the Bible has been described as a land "flowing with milk and honey." To the very young, in Sunday school, this is an extremely real picture.

However, as we grow older, we realize the Bible was written by many Godly and gifted men who often wrote symbolically of things as they saw them and as they were impressed by both the appearance and the thoughts which surround them.

The ancient Hebrews must have been very strong people of great endurance to become accustomed to the great and many contrasts in climate of their area.

In the Book of Deuteronomy we find part of the land spoken of as, "great and terrible wilderness." The intense heat and dryness is described which leads us to believe any tree offering a tiny bit of shade was, indeed, a blessing!

Then again, in the same Book of Deuteronomy, we find described a part of the land with brooks of water, wheat and barley fields, vines, fig trees, and pomegranates.

From the Old Testament we also learn those early sons of Israel were extremely familiar with the perfume of the orange blossoms, the fig, the pomegranate, and the apricot trees. There were the woody fragrances of the stately cedar of Lebanon and many times there is mentioned the wild flowers of the fields.

In the New Testament we are told our Master saw and loved the wild flowers of Palestine.

Today botanists tell us there are more specimens of wild flowers to be found in a part of the Holy Land than anywere else in all the world!

Those who visit the Holy Land today are overwhelmed at the brilliance and profusion of the lovely poppies, cyclamen, phlox, and oleander. How wonderful it is to picture Jesus walking over those beautiful hills in the springtime, and to know He, too, enjoyed the fragrnance wafted to Him from the many blossoms.

After my lectures, there are frequent inquiries pertaining to the perfumes mentioned in the Bible.

It has been extremely difficult to find some of these fragrances because they are not like the popular scents of today. Most of them are of the "resin" type, which means they are the sap or juice trapped from shrubs or trees. They are all exotic and extremely interesting. The research concerning these ancient perfumes of the Bible proved to be most exciting. Let us investigate a few.

Aloes was a perfume mentioned both in the Old and New Testaments. Numbers 24:6 tells us, "As the valleys are they spread forth, as gardens by the river's side, as the trees of lign aloes which the Lord hath planted." And we find, "thy God, hath anointed thee with the oil of gladness above thy fellows. All thy garments smell of myrrh, and aloes, ...out of the ivory palaces." Psalm 45:7,8.

The aloes mentioned in the Old Testament is also known as Eaglewood (Latin) and was first found in Asia. Agallochum is the heartwood of this tree. In the inner trunk of the center of the tree can be found a dark-colored highly fragrant substance. This fragrance does not carry down to the smaller branches. This soft inner wood could be quite easily molded so it was used extensively as mountings or settings for precious stones in Bible times.

To this day, in the East, there is a belief that this tree actually descended to man from the Garden of Eden. According to their story, Adam was supposed to have brought a tiny shoot from the Garden and planted it in the land where he lived and died. For that reason, the tree is known to many as the "Shoot of Paradise" or "Paradise Wood."

The aloes, mentioned in the New Testament, appears to be the true aloe which is the native of Socotra, an island in the Indian Ocean. This is a succulent plant, not a tree. A bright violet liquid can be pressed from the leaves of this plant. The substance can be dissolved in water and was used as one of the ingredients in sweet-smelling incenses and also in purifying the bodies of the departed.

The aloes used in perfumes is not the bitter aloes known

in medicine as a purgative. It is the wood from the aloes tree which grows chiefly in Bengal, Burma and Java. There they distill oil from the wood and it is known as "agar-attar." This has an odor very similar to sandalwood.

It is of value to know only the aloes trees which have been attacked by a certain disease prove to be aromatic. Therefore, when the trees are cut down, they are immersed in water because the wood that floats is discarded and the wood that sinks yields the prized perfume oil. Evidently this is the reason the Chinese call aloes "fragrance sinking under water."

From the extremely sweet-scented timber of the almug tree were made the pillars for the house of the Lord (I Kings 10:11, 12.)

This fragrant wood today is known as "Red Sandalwood."

In Biblical times, musical instruments, harps and psalteries were also made from this wood.

Sandalwood chips were frequently strewn upon couches and used to perfume dwelling houses. This was done to mask, or cover, the stench of the dwellings and meeting places of those days. Remember there was no sanitation then, no running water, and which seemed to bother those of that age very little. There is no question in my mind that perfume did worlds to make life bearable and definitely more pleasant.

Today sandalwood is still used in the incense burned in temples during religious rites because inhaling this delightful fragrance makes one generous at heart and the collection plate much heavier.

It is thought by those who have made an extensive study of such plants that the anise, which is mentioned in Matthew 23:23, must have been dill. We are quite familiar with both as a seasoning for our food, but in Biblical times both were also used as a perfume. We are told, "It is a carminative and yields essential oil." Even today the dill fruit is used as a universal medicine in India. In ancient times it was only grown by the Greeks and the Romans.

"I gave a sweet smell like cinnamon and Aspalathus" is found in Eccleisiasticus 24:15 in praise of wisdom. Because all wisdom came from the Lord, it was likened to the plants giving forth the sweetest of perfumes. This plant is a shrub which yields lignum rhodianum, an essence of delightful fragrance adored by the ancients. This essence was used by them in preparation of an ointment for thickening the hair and beards of men, although it was also applied to the bodies of all who could afford it for the simple joy its fragrance gave them.

Balm, which is balsam, is mentioned many times in the Bible. This is a gum or thickened juice extracted from the balsam tree--especially those of Judea.

The Jews believed it had been planted on the plains of Jericho by King Solomon. History tells us the very first roots were brought to him by the Queen of Sheba when she paid him a royal visit. We find this in Kings 10:10. The plant is used in medecines, but within the pulpy nut a seed is found which is most fragrant.

The balm mentioned in Genesis 37:25 appears to be the false balm of Gilead which was an evergreen shrub found growing in quantity on the plains near the Dead Sea. The lovely white blossoms of this shrub turn into green fruit, resembling our apples. These fruits are picked before they completely ripen, and from them a sweet oil is extracted. The gum resin from the bark is most fragrant.

In Genesis 2:8, 9, 12 we read of bdellium. This is an odoriferous gum from a tree. History tells us when the gum was removed from the bark of the tree, the pieces would soon harden. In hardening they became transparent and like bits of wax and, in fact, looked like pearls. The women of ancient Egypt carried little pouches filled with these pieces of bdellium. They not only looked like pearls, but gave off a most delightful fragrance as well. We find still another mention of this in the book of Numbers 11:6,7.

Parkinson, who was a 16th century English herbalist, writes that when the bark of this tree is incised, the gum that oozes out is the size of a white olive, noted for its fragrance.

Calamus, so often mentioned in the Bible, is obtained from the rhizome or the root-stock of the sweet flag. It has a terrific sweetening quality when blended with other fragrances.

In the Song of Solomon 1:14 and 4:13 we read of camphire which is one of the earliest known spices and perfumes. It is also known as the Henna Flower, even at times is called the Cypress Flower. The ancient women used it to stain the palms of their hands and the soles of their feet.

Its fragrant flowers were worn as chaplets around the neck and used to decorate and scent their homes.

In later years it was found to be a most effective check to excessive perspiration.

We know camphire has been traded for centuries by evidence found in tombs. The nails of mummies which had been dyed with the camphire, or henna, retained the stain after being entombed for 3,000 years! The leaves of the plant are pounded to give off the dye and the fragrance.

In the formula found in Exodus 30:20, 23, 24, 25 we read of cassia. This is a fragrant tree which resembles the cinnamon, but is less delicate in taste and scent. Its fragrance is extremely sweet; to wear it as a straight perfume is almost too overpowering, but when blended with softer floral fragrances it is delightful! We find this mentioned also in Ezekial 27:3, 19.

Cedar was another favorite fragrance of the ancients. In Numbers 19: 1, 2, 6 we are told it was one of the shrubs used to cover the disagreeable odor of the burned offerings. The cedar is a strongly rooted tree. Its wood has been used extensively down through the ages to make scented chests. The fruit, when burned, gives off a pleasant odor.

Cinnamon is mentioned in Revelation 18:11, 12, 13, 15 and also in Exodus 30, 23, 25, 26. The perfume is the oil extracted from the bark. In the Biblical times sweet cinnamon was the symbol of understanding and so was used for this reason in all anointing oils.

We are told cummin had value in Isaiah 28:26, 27 and also in Matthew 23:23. Centuries ago, we are told, Persia raised this plant for seeds which brought a very good price in trade. It is similar in shape to the fennel and contains a very valuable oil which is strongly aromatic. When used as a spice, it is crushed and mixed with bread and meat, as in a meat loaf. It is interesting to know that Kammon, which is a village near Acre, has taken its name from the Hebrew word for this particular flower because of the sharp smell abounding there. The fruit, when ripe, is rounded and quite dry. It is harvested by beating the stalks with a rod even to this day and age. In this way the small and tender seeds are preserved, but would otherwise be ruined if other methods of threshing were employed.

The fir, or pine, and its costly resins were well known to the ancients. The essential oil, which we know as turpentine, was used in the Greek Fire of medieval warfare. We find the fir mentioned in Isaiah 60:13, 14 as well as I Kings 5:10. The fir oils were used often as a base for the perfumes burned in lamps in the Holy Land, Persia and Arabia.

Even though it does not yield a perfume, let us hesitate a moment to consider the flax. It was from flax that the linen was woven which was used by the Egyptians and the Jews, after dipping it into the perfume oils, to wrap the bodies of their dead. It is mentioned in Joshua 2:3, 4, 6 and Luke 23:50, 52. The soft linen woven from the flax retained the perfume oils in sufficient quantity to completely seal the bodies. Thus, they have been preserved all through these many years.

The minute perfume is mentioned, everyone recalls frankincense. This is mentioned at various times in the Bible, but I think the lovliest story is found in Matthew 2:10, 11.

When the Magi followed the Star of the East to the birthplace of Christ, gold, frankincense and myrrh were the gifts they brought to him. It is most interesting to realize that even today on the feast of the Epiphany, in commemoration of the Magian gifts, little silken bags containing gold, frankincense and myrrh are still placed on the alms dish of the Chapel Royal in London by the Lord Chamberlin of the Queen of England.

These particular gifts were brought to the Christ child because they were the most valuable things in the world of that day.

The first was gold given in token to kingship, the frankincense was the symbol of holiness, while myrrh symbolized the the suffering the newborn infant would have to endure in His life here on Earth.

The frankincense tree is really a thing of great beauty. It bears lovely clear green leaves which resemble the mountain ash, and it has pretty star-shaped flowers which are a beautiful pink with lemon centers. The wood is the heavy, hard durable type and has been used for many things down through the ages. We are told at the end of February the bark is cut and a thin layer is peeled off. One month later this process is repeated and then the juice, or resin, flows out from the inner wood.

When this becomes hardened, it is brittle, glittering and very bitter to the taste. We are told it is recognized as the finest burning resin, or incense, in the world; hence its name "frank-incense," or "free-lighting."

This fragrance has always been used for services in the Holy Temple, and was one of the resins used for fumigation to mask the odor when beasts were offered for sacrafice.

In Biblical times Pure Frankincense represented praise, heart adoration, and, therefore, was used extensively in temples.

In Exodus 30:34, 35, 36 we find galbanum mentioned, which was another ingredient in the incense burned at the golden altar in the Holy Place. This is the gum, or resin, taken from the plant Ferula Galbanifula.

The gum is collected by cutting the young stem a few inches above the ground. Soon a milky juice flows out which soon hardens.

This gives off a lovely pungent odor as it burns. Today it is used in varnish as well as in the medical world.

Green bay tree, mentioned in Psalm 37:35, 36, is seldom thought of in connection with perfume. However, when its evergreen leaves are broken, they emit a very sweet scent. This has furnished an extract used for centuries by the Orientals in the making of perfume. The root and the bark supply a valuable medicine. Then, too, in the ancient Olympic games the victorious contestant was awarded a chaplet of bay leaves which was placed on his brow with great ceremony.

The hemlock mentioned in Hosea 10:2, 3, 4, and Amos 6:12, is seldom, if ever, used as a perfume today. However, I have found it does have a great aromatic value which has been passed down through the ages.

It was first brought to my attention during a visit at Cooks' Forest, Pennsylvania, when I noticed the small general store heavily decorated with boughs of hemlock. Being curious, I asked them why they had decorated in this manner. I shall never forget the look of utter astonishment on the face of the young boy, who was a native of the forest, when he explained it not only attracted the flies but it also took away the odor of all tobacco smoke in only a matter of moments! Since then I have frequently used hemlock boughs with my floral decorations when visited by those who smoke, and found it works wonderfully!

It is believed when lilies are mentioned in Ecclesiasticus

50:8 that the lily "by the rivers of water" was really an Iris. The word Iris comes from the Greek meaning a rainbow. The fragrance of this flower is so very faint it can only be detected immediately upon the opening of the flower. However, some Iris roots bear a very delicate odor.

Through the ages they were dug up in the autumn, slowly dried in the shade and placed in linen chests by many to add a scent of perfume. The ancient woman also threaded some of the small roots on linen thread when she hung them among her family's clothing.

The Lilies of the Field mentioned in Luke 12:27 is now known to be the Anemone, or Windflower. These flowers have a very delicate fragrance which I have found to be delightful when blended with the fragrance of other early spring flowers.

The lily mentioned in Song of Solomon 5:13 was especially favored as a symbol of lovliness. It grew in the gardens of King Solomon, the regal botanist, who greatly admired its striking beauty and color of "glowing flame." I have always thought of this as our "Tiger Lily." I have never found a perfume extracted from this lovely flower.

In Biblical times the locusts, the fruit of the carob tree, mentioned in Matthew 3:1, 2, 4, 5, 6, were spoken of as a food. However, especially in later years, the flower was highly prized for its very delicate but penetrating perfume. This fragrance always reminds me of our sweet-peas. I find the fragrance of locust blossoms lends a delightful depth to my Oriental blend. It is a scent that seems to remind one always of something in the past: sweet, illusive, but definitely pleasant to remember!

The mastick tree mentioned in Susanna 51:54 gave the ancients an entirely different type of perfume. We are told "Women living in harems use the resin obtained from the bark by incision. They chew it to sweeten their breath; hence this mastication gives the tree its name."

Oil obtained by pressing the berries of the mastick tree has always been used by the Arabs for a food as well as illumination. This gave off a lovely soft fragrance as it burned.

The twigs from this tree, which are light and very flexible, have been used in ancient times to make nice smelling baskets.

In the New Testament, Luke 11:42, mint is mentioned. This herb had many uses. I was delighted when I read, "The Jews strewed the floors of their synagogues with mint so that its perfume scented the place at each footfall."

The myrrh mentioned in the Old Testament, Genesis 37:25, 26, 27, seems to be a small shrub or rock plant. Some writers call it the "Rockrose." At any rate, this plant provided a very sweet-smelling gum for all its parts and the peasants still gather it as they did in ancient times.

Their method of gathering this gum is quite unusual; they

use a small stick wound around with a soft cloth, and on a calm day, they carefully wipe the sweet substance from the shrub and round it into balls. Later this is firmly pressed into cakes and used for perfume. It has a very pungent fragrance, but serves as a lovely mellow foundation when blended with lighter and gayer oils.

The myrrh of the New Testament, Matthew 2:11 and John 19:39, 40, seems to be quite a different type of plant. It seems to have been a large bush or tree. The perfume odor is much the same, but the extraction certainly was a lot more simple!

The trunk of this tree is large and has numerous knotted branches. When the bark is pierced a thick white gum appears, which hardens and turns reddish upon exposure to the air. It is this aromatic gum which has been known to be priceless through the ages. It is used for a spice, as medicine and perfume.

Myrtle has always been famous in the world of scent. Its very name is taken from the Greek word meaning "perfume." It is mentioned several times in the Bible; Zechariah 1:7, 8 and Isaiah 55:13. It's a large evergreen shrub which grows into a large tree, sometimes over eighteen feet high. The Jews have always used it to adorn their booths at the Feast of Tabernacles. It has beautiful white flowers, and the perfume from the blossoms has always been considered more exquisite than the perfume of a rose! Even today in Italy the leaves are used as a spice and in Syria all parts of the plant are dried for perfume.

Onycha, mentioned in Exodus 30:34, 35, 36 is a most interesting fragrance and a still more interesting plant.

It is a rockrose which produces a gum that is known as "labdanum." The blossoms are about three inches across, white with at the base of each petal a blotch of brilliant scarlet-rose which deepens into black. In Greek the word "onchya" means "fingernail." The blotch of color in each petal looks exactly like a brightly painted red fingernail.

Late in the year a soft glutinous resin, which is highly aromatic and delightfully fragrant, exudes from the leaves and stems. In Biblical times, this resin was highly regarded and extremely expensive as a perfume.

The Bible mentions many roses. In Ecclesiasticus 39:13, 14, 15 we believe the rose to be our oleander. Students declare the "rose of the waterbooks," the "rhododendron," or the "Rose Tree" of the Greeks to be the oleander as we know it today. The fragrance is most delicate and seldom used as a perfume.

The rose spoken of in II Esdras 2:10, 18, 19 was the Phoenician rose which grows in a large bush. This is a cluster rose with a very sweet perfume.

The rose which is mentioned in Isaiah 35:1, 2 is far different. In this instance, we are told, the Hebrew word translated as "rose" indicates a plant with a bulbous root, so it couldn't possibly be a rose bush. Those who have studied the subject at great length assure us this rose was really the Narcissus. As proof, they tell us the earliest Chaldean paraphrase of the Holy Bible gives the Hebrew word "narkom" which means Narcissus.

In the Holy Land, we are told, this narcissus-rose is a glowing, dazzling golden yellow such as we do not see in the cooler climates. Its perfume is heavenly! My "Touch of Spring" brings this fragrance blended with the Jasmine bud of Egypt plus our own Southern Magnolia, from the tree. It is truly a spring fragrance which quickens the pulse and causes spring with all its love and romance to burst forth from our very souls!

Another rose of the Bible is the Rose of Sharon spoken of in the Song of Solomon 2:1, 2. We are told this is not a shrub or vine but a bulb growing plant known as our tulip. The fragrance of this flower is one of freshness rather than perfume.

Rue is mentioned only once in the Bible, Luke 11:42, 43. It has been very valuable through the ages as a herb. The specific name, we are told, is "graveolen," which means "strong-smelling." In ancient times it was gathered for use as a disinfectant, scattered in the courts to protect officials from the dreaded prison fever. Suggestions that rue gave forth a perfume is quite a "strong" statement. It did, however, serve its purpose well in overcoming the stenches which prevailed in these ancient times.

In Biblical times, saffron, Song of Solomon 4:13, 14, 15, 16, was known to be one of the most important of all the condiments and the sweetest of all perfumes. The only disadvantage was it left a yellow stain.

It is the stigmas of this flower which are valuable. They are a vivid orange in color, dry, narrow and almost threadlike, but emit a most peculiar, sweet and aromatic odor. The flower itself is a delicate lavender in color and the fragrance is light and most pleasing. One more thing of interest: "One grain of commercial saffron contains the stigmas of nine flowers; some 4,000 blossoms are required to make one ounce."

Saffron was considered of such great importance in the Middle Ages that during the 15th century, in Nuremberg, men were burned at the stake and even buried alive for the crime of adulterating it.

We find the shittah tree mentioned only in Isaiah 41:18, 19, 20. The word "shittah," which is the plural form in Hebrew, is mentioned many times. We know this tree to be the Acacia. It has soft green leaves and lovely yellow flowers simply bursting with sweet fragrance. Much is written in both the Bible and in history regarding the Acacia. It is believed the Ark was made of its wood. The tree also exudes a resin which is known as "Gum Arabic" and "Gum Senegal," both well known today.

Spices are mentioned countless times throughout the Bible. We will dwell only on the one spoken of in II Chronicles 9:1, 5, for these are the spices used for the holy incense in the Tabernacle. I have found, "In the English Bibles the word for the astragal plant is inaccurately translated "spices." Its products are not proper spices. It is the gum tragacanth of commerce, and the plant is known as wild tragacanth or thorny astragal, the "necoth" of the Bible.

It seems the Biblical species is dwarf and simply covered with prickly thorns, spines, and edges as sharp as razors! On this plant it is the thorn that exudes the precious resin and only during the hours the plant is bathed in sunshine. The ancients rubbed a ball of cotton over the plant, thus collecting the accumulated lumps of gum or resin.

We find the spikenard mentioned in Song of Solomon 1:12 and also in Mark 14:3, 4, 5, 6.

I found the Indian word for spikenard is "tamul," meaning an herb with a most agreeable perfume. King Solomon rejoiced in its fragrance. Its Latin name is "Nardostachys" which means "ear of wheat," evidently, in reference to the shape of the flowerets. The Indian name "Jatamansi" relates to the shaggy hair, or "ermine tail" covering the stems.

The ancients exported large amounts of this fragrance in the form of ointments made from this precious plant. It was known as "singul Hindi" or "Indian Spike."

The Romans used it for anointing the head. The ointment was a rich rosy red and had a sweet fragrance. From the lower hairy stems comes the most exquisite perfume which is obtained by simply tying them together by the roots.

Can you possibly imagine one little plant having all this most interesting history? There is still more!

It is said, "so very expensive was the perfume (from this precious little plant) after its long journey from northern India to Palestine, that one pound of it cost 300 denarii. The denarius was the equivalent to a laborer's daily wage."

So priceless was this perfume that it was transported on camel-back, sealed in alabaster boxes to preserve the essential perfume oils.

Stacte (Storax) is a resin which exudes from certain trees, such as the myrrh tree, and was a part of the perfume formula given in Exodus 30:34, 35. We are told the Hebrew word "nataph" really means a "liquid drop." It seems when incisions were made in the branches, the resin flowed out in a liquid state. The ancients gathered it in reeds, which has given the resin the name of "storax kalamites." When the resins hardened, the stacte (or storax) was then scraped off in irregular compact masses which were always interspersed with smaller drops known as "tears." These drops, or tears, contained resin and benzoic acid as well. The ancients were delighted to learn they would dissolve in wine. Could this possibly have been the beginning of perfume with alcohol? Records do not tell us whether they wore this perfume dissolved in wine or drank it!

We find in very early days in England this same storax was imported and used in perfumed pomades. Even today the Roman

Catholic Church uses it in their incense.

The history of sweet cane (sugar cane) as mentioned in Isaiah 43:23, 24 proved to be most interesting to me.

In fact, I almost passed it by until I noticed a footnote which read, "Sweet cane (calamus)." The mere word calamus started me on a real search for I have found this fragrance to be one of the sweetest and most interesting of all the scents I have managed to collect from the Holy Land. I found the Biblical "Canes" were sweet to both taste and smell. The aromatic odors were mentioned in Exodus 30:23, 24; I Kings 10:10; Song of Solomon 4:14; Jeremiah 6:20; and Ezekial 27:19 under the heading of "Calamus," "Sweet Cane" or even "Spices."

We find the mention of thyine wood in Revelation 18:12, 14, 15, 17. As we read of the fall of Babylon, we note this wood listed among its most valuable and costly possessions. In the flourishing days of the Roman Empire this was called "Citronwood," and used extensively for fancy and ornamental woodwork. It was such a durable wood it was absolutely priceless! Not one bit was ever wasted, for all the tiny scraps left from building were used as incense. The fragrance it gave off was delightful to the ancients.

The resin exuded from this precious tree was a whitish-yellow often called "Sanderach," which is frequently mentioned in ancient history. The resin was so completely free from impurities that the ancients used it in the preparation of all

their parchment of that day. In later years this was rediscovered--
they found when the resin was dissolved in wine it made a marvelous clear varnish.

The turpentine tree mentioned in Ecclesiaticus 24:16, is really the "teil tree" (Isaiah 6:13). This is a lonely sort of tree, since it never found in groups or woods. The ancients found, when the bark was cut, a liquid resin fairly gushed forth! They were pleasantly surprised to find it had a most agreeable perfume which rather resembles the jessamine flower (the State Flower of South Carolina.)

The resin from this tree has many of the properties of our turpentine, but certainly does not have the harsh aroma of the turpentine obtained from our pines. Once again this resin would harden when allowed to stand in the open air for a time, and it burned, giving off a lovely glow and fragrance.

We are all familiar with water lily, which is spoken of in I Kings 7:19, 22, 23, 26. The perfume from its lovely flowers is extremely unusual and very heavy. This scent is favored highly by the Egyptians. It is also used by those in Arabia and China. I have found it to be most delightful in combination with some of the lighter Oriental perfumes, and have used it frequently.

In many passages of the Bible, we find reference to the wide use of perfume in those days.

Babylon was the chief perfume market and was famous for its

choice scents which were kept closely guarded in flasks of alabaster.

The Babylonians and Syrians did not regard their perfumes as either sacred or medicinal--just precious! At times of festivals they went out and used perfumes most recklessly because it gave them a feeling of great joy and luxury to do so.

As an example, we are told, a series of games were held at Daphne during which each one who entered the area was liberally sprinkled with at least fifteen different kinds of perfume. Some of the scents were taken from the lily, amaracus, spikenard, saffron and cinnamon.

The perfumes were carried in golden watering pots in the arms of two hundred women chosen for their great, and natural, beauty. In Babylon, King Hammurabi passed a very special edict commanding every person in that area, or in his kingdom, to bathe himself every single day in perfumes.

One would think the King had a corner on the perfume market, eh? If so, it was not recorded in history!

Needless to say, this edict accomplished one thing: it did wonders to cover the body odors which prevailed during those days.

Chapter 14

Herbs

In Webster's Dictionary, we find the word "Herb" defined as, "A seed plant which does not develop woody persistent tissues, as that of a shrub or tree, but is more or less soft or succulent; specifically, one used for medicinal purposes, or for its sweet scent or flavor." The example given is "Grasses."

Were this entirely true, bay, rosemary, peppers, mutmeg and cinnamon would not be considered herbs. Therefore, I prefer to think of an herb as a multi-purpose plant because it has been used for centuries for healing the sick, for flavoring food, for making perfumes and many other forms of scent.

We are told a 1,000 years ago an herb garden was really an apothecary's shop. Today it is quite different. Herbs can be grown by anyone who wishes to enjoy their sweet-scented ingredients either in an enchanting garden, a window box or on a kitchen windowsill. They are eager to grow and bloom for all. The most wonderful thing of all is that they require very little care or tending.

Their matchless fragrances, brought out by the heat of a summer day, or the cool dampness of early evening, thrills an observant and sympathetic gardener right to their toes!

Many herbs are still used in the healing arts. For those who are allergic to drugs, chemicals and synthetics there is no choice; herbs are the only answer to their health problems.

Naturally, my first interest in herbs pertains to their scent, their perfume, their fragrance given off so freely to anyone who cares enough to work with them.

The one discovery I have made in this field of study follows as this: There is nothing in all the world alcohol can do that herbs cannot do much better!

Chemists may doubt this but I have proven it to my own satisfaction. You see, this bit of research was prompted by the fact that my olfactory nerve, my sense of smell, is completely dulled instantly by the fumes of alcohol. Then I begin to sneeze like crazy!

Therefore, I had no choice, I could not use alcohol in my lab. Very soon, however, I found the precious herbs most eager, willing and able to substitute. The happiest discovery of all was when I learned the essences extracted from herbs did not evaporate, they remain constant in both form and fragrance when extracted properly.

Today, herbs are once again coming back into their own. Many books have been written telling us of their endless value. The list I have before me holds great promise such as the art of cooking with herbs, an herbal book for the dog in health and sickness, herbs for the farm and stable, while one even explains that herbs are the fountain of youth! What more could one possibly ask of an herb?

The aromatic herbs are delightful in the home. When arrangements are made of the lovely silver sage leaves, both green and purple basil with, perhaps, a bit of wormwood, it becomes a glorious creation. Within moments the room will be filled with the blended fragrances--all most pleasing. When included in a bouquet, herbs also serve another purpose: they do wonders in keeping the water sweet for a much longer period of time.

Sprigs of lavender in the bedroom, bath, linen closet or storage chests saturate all they touch with a lovely clear, clean scent welcome to all.

Several of my friends, who belong to the Herb Society, prepare Advent wreaths for their churches. These wreaths are a delightful combination of both the green and dried herbs. When the green herbs of the rosemary and bay are used together with sage and thyme, they tell me, the church is immediately filled with a lovely fragrance entirely different from either floral or evergreen arrangements.

We are told the first wives, who left for the New World in sailing ships, managed to tuck away in their trunks many of the seeds, crushed leaves and even a few roots of the herbs they had been taught since childhood to depend upon for many and varied uses. It was these tiny packets of seeds which started many of the old fashioned herbs we now have in this country. Bless them for their thougtfulness!

Before they came to America, the wives had learned to crush the dried herbs and had made "Dry Perfume" which later became known as "Sachets." In a very old book, I found the following formula:

"Herbal Dry Perfume"

```
Calamus root........................1 lb.
Caraway.............................1/2 lb.
Lavender............................1 lb.
Marjoram............................1/2 lb.
Musk................................30 grains
Cloves..............................2-3/4 oz.
Peppermint..........................1/2 lb.
Rose leaves.........................1 lb.
Rosemary............................3-1/2 oz.
Thyme...............................1/2 lb.
```

I have never tried duplicating this formula but doesn't it sound delicious?

Today herbology is an accepted science. Scattered over America are thousands of men who make their living collecting valuable wild herbs. They call themselves "herbologists," "leaf peddlers" or "wildcrafters." They follow the flowering plants north in the spring to the Canadien border and then travel south in the autumn to collect the valuable rootstocks.

They, actually, perform a most useful function. Without them, druggists would be unable to fill some of their prescriptions and many businesses would be at a standstill.

At one time the American druggists had to depend largely on Europe and Asia for most of their herbs. However, the war heavily damaged the foreign sources of supply, and today a great many roots and herbs found in this country are bringing the highest prices in almost thirty years! With the building of more cities, more super-highways, year after year the supply of wild herbs becomes more and more scarce. For that reason, those who enjoy the flavor of herbs in their food, and especially those who wish to use herbs for their fragrance, find it necessary to establish their own herb gardens. It takes time, at first, but the dividends are high, and the pleasures unlimited.

Chapter 15

Other Sources Of Fragrance

"Acacie" or "Acacia"--Latin: Acacia Farnesiana. French: Cassie. German: Acacie.

We are told the flowers of the Acacia Farnesiana are a native of the East Indies flourishing farther north than the other varieties. At this time it is cultivated to great extent in southern France for the delightful odor which resembles that of violets, but is far more intense. It is only the flowers which are collected and from which the fragrance is extracted.

The Acacia Biflors and the Acacia Hastulata are famous the world over for their scents. The ancients gathered the flowers after sundown and macerated them in olive oil. The most famous perfume of Australia, known as "Wattle Blossom" perfume, has this fragrance as the heavy note.

The plant which is generally but falsely called Acacia in this country is really Robinia Pseudoacacia. This plant also bears very fragrant flowers from which could be made perfume by employing the same methods.

It is amazing to realize that the Black Currant, Ribes Niger, contains in its flowers an odor so closely resembling the Acacia fragrance that some have made perfume of it which fooled even the experts! The latter is used mostly in solid perfumes and they have the grace to call it "Oil of Cassie." The Black Currant will grow in the United States--even the northern part--

but the Acacia Farnesina could never stand our rugged winters.

The Acacia can be blended with bergamot, bois de rose, geranium, jasmine or orange blossoms to advantage.

Still another perfume of note is the famous and very old bergamot.

The bergamot is the fruit of a tree belonging to the Order of Aurantiaccae cultivated first in Calabria. The tree is unknown in a wild state. The golden-yellow fruits, which resemble a lemon in shape, have a very bitter, acid taste.

The fragrance is obtained from the rind of the small and very bitter orange. We are told it is now grown in large groves as Messina "by the blue Sicilian sea." This oil is used largely in the manufacturing of fine perfumes and soaps exported chiefly from Messina and Palermo.

In days gone by snuff was always scented with bergamot.

In the works of Cowper we read, "Give the nose its bergamot..." which, of course, meant snuff scented with the fragrance of bergamot.

The name, bergamot, we are told was taken from "Bergamo," which was a famous old rustic dance sometimes called the "Bergomask." This is mentioned in Shakespeare's "A Midsummer Night's Dream."

Today we find this fragrance used mostly in toilet preparations for men--the most expensive type.

Candleberry, which is also known as "Candleberry Myrtle"

or "Wax Myrtle," is a shrub which grows from four to eighteen feet high. It is quite common in North America where candles are made from its drupes or berries. The berries are about the size of peppercorns and all covered with a very fragrant greenish-white wax popularly known as "Bayberry" tallow.

The wax is collected by boiling the drupes in water and skimming off the surface, exactly as the ancient Egyptian Priests made their first perfume!

A bushel of berries yields from four to five pounds of wax.

Another plant belonging to the same genus is the sweetgale which grows abundantly in bogs and the marshes of Scotland. This latter is a small shrub with leaves somewhat like the myrtle or willow, of a very fragrant odor but very little taste which also yields an important essential oil by distillation.

For many years "Carnation Perfume" was made from cloves!

If properly extracted, blended and aged, it is most difficult to distinguish between the two scents.

Cloves, as we know them, are the unopened flowers of an evergreen tree which grows on the Indian Ocean island of Zanzibar. The buds are picked by hand twice a year. I was amazed to learn some of the trees are over century old and are still blooming!

Did you know the dear little forget-me-not is not only known for its delightfully delicate perfume but is also well known as an ancient herb?

This is true! The ancient Egyptians regarded its juice as a very valuable remedy for weak eyes. The ancient Priests advisee wet packs of the juice to be frequently used when the eyesight appeared to be failing.

I have found this delicate fragrance, extracted by my enzyme brew method, to be a perfectly wonderful "extender."

As an example, the fragrance of the Tuberose is most potent, often overpowering, and termed as "strong" by those who inhale its scent. In spite of this, Tuberose projects a most important emotional attitude desired by many. To bring this in a suitable solution to reach all desiring it, I have found it most advantageous to add a bit of fragrance extracted from the forget-me-not to the potent Tuberose. The result? The answer to a most important problem!

The lovely fragrance of fougere has floated in and out of the history of perfume many, many times. During some periods it was most popular, especially in soaps. Then again, when sweeter perfumes were all the rage, the fragrance of fougere would be forgotten for a while.

Still, down through the centuries, it has never disappeared entirely. Fougere, you see, is a fern.

The dictionary tells us, "Fern--any of a large order (Filicales) of pteridophytic seedless plants, like seed plants in being differentiated into root, stem, and leaves (fronds) and having vascular tissue, but reproducing by means of asexual spores that develop into thalloid gametophytes."

Sounds impressive, doesn't it? But quite confusing. A bit further on we read, "Fern seed--the dustlike asexual spores of ferns, formerly taken for seeds, and reputed to render one invisible." Now this may explain the disappearance of the fern itself from time to time! Who knows?

Quite clearly, early on in my research, I found the fragrance of _some_ fougere, or ferns, greatly interested me, while the fragrance of others actually annoyed me. In time I found the essence extracted from a fern, or fougere, which had grown in the heat of the sun was much stronger and not as pleasing as the fragrance extracted from a fern which had spent its life in a shady dell beside a bubbling brook.

It is the fragrance extracted from the latter fougere which gives a woodsy perfume its firm foundation and its strength of character.

When walking in the cool of a forest or the shade of a wood, it is the fragrance of the fern growing there which urges you to penetrate deeper into the woods. The same beckoning gesture continues on when the fougere is added to a perfume. It is cool, inviting, and always promising.

The famous Geranium oil is always extracted from the leaves of the plant.

This can be taken from the Martha Washington Geranium, the Rose-leaf Geranium, or the old fashioned Sweet-scented Geranium.

All have a similar fragrance--sharp and distinctive--but each type has a little different reaction on the one who sniffs it!

There are similar fragrances to be found in a few other plants and flowers also.

It is said by many that the Bois de Rose, the Sweet Briar and a few other roses have the "Geranium Type" scent. Some say it may be found in the Eglantine.

My memories, however, of Geranium restrain me from believing this. Those memories are very near and dear to my heart. They force me to believe there is actually nothing in all the world to compare to the fragrance obtained from the leaves of the Geranium.

My mother loved the fragrance given to us by these precious leaves. From her plants, she would pluck a leaf--at its prime--place it with her linens, and especially her dish towels. When that door of the linen closet was opened, when that drawer in which her tea towels were kept was pulled forth, the fragrance wafted up was simply out of this world.

I remember well, when she made crabapple jelly, she always carefully laid a geranium leaf on top just before applying the sealing wax. The taste, the fragrance, made her jelly a delicacy long to be remembered.

Another cherished memory is that of "Geranium Mountain." Many years ago, on my first visit to California, friends took me to see this sight. I often wonder if it could still be there, just outside the city of Pasadena.

I shall never forget my reaction as we neared this mountain (actually one of the smaller foothills) which appeared to be on fire! The blaze-red, a color produced by the noonday sun, was breathtaking.

Then our friends told us that many years before an easterner had planted some Geranium slips on the side of the mountain, hoping they might grow. Well, they certainly did, for the mountain was blanketed with a carpet of its blossoms! You see, there it grew--and bloomed--twelve months of the year, for it was always shielded and never touched by frost even when snow came on top of the mountain.

I remember leaving the car to investigate this plant more closely. To my amazement, it had many arms, many limbs much larger than mine. They were covered by a haeavy bark like a tree.

Surely this plant had found its place in the sun. In remembering it all these years, I always think what a lovely living monument that flaming mountain has become to the courageous easterner who planted its first slip.

Iris--the beautiful Florentine Sword-Lily, which is really the Iris Florentine, often grows wild in Italy. It is largely cultivated--especially in America--where we have Iris Societies organized just for the study and cultivation of this lovely flower.

It has a creeping root-stock which is covered with a brown bark. When this covering, or bark, is peeled off exposing the fresh root, it becomes the world famous "Orris Root" known so well in the perfume and cosmetic world.

Orris Root occurs in commerce in whitish pieces which are sometimes forked; the surface is knotty and the size may reach the thickness of a thumb and the length of a finger.

When fresh, the roots have a disagreeable odor, but upon drying they attain an odor which may be said to resemble that of a violet. Upon comparing the two fragrances immediately, a considerable difference is quite perceptible, even to the untrained olfactory sense. Still, there is a similarity between the two, vague as it may be.

Orris Root should be as fresh as possible; this may be recognized by its roughness, the great weight, and the white, not yellow, color when the root is cut open.

This most valuable root is frequently used for sachets and is also a most excellent vegetable fixative when used to hold other fragrances.

You have just learned one of the most carefully guarded secrets of the perfume world. Isn't it fun? From now on, I hope, the Iris Roots will hold a greater meaning for you.

One of the oldest and most precious of all fragrances is that of the Jasmine blossom. This is such a delicate fragrance it has been absolutely impossible for chemists to produce a synthetic substitute.

The ancient Hindu poets have written many lovely poems in honor of this fragrant flower in which they have referred to it as the "Moonlight of the Grove." Throughout history, it has symbolized "divine hope" or "felicity." The Chinese think of it as the symbo of "purity and sweetness of womanhood."

It is interesting to know the Chinese selected this flower as very near and dear to their hearts by using the buds to perfume their ritual tea. Its flavor is truly delightful!

The ancient Arabian merchants carried "cuts" and "slips" of the Jasmine shrub into Egypt where it was planted and cultivated. There it came to be called "sanbac." I believe there is nothing as lovely, as delicate, or as lifting, when added to a blend of perfume oils, as the gentle fragrance extracted from the Jasmine bud which grows beside the Nile River in Egypt.

However, Grasse furnishes us with a better perfume extracted from the full blown blossom. It seems the climate in southern France gives the fragrance a strength and power which is necessary for the heavier blends.

Not far from the coast, near Grasse, France, grow large fields of the precious Jasmine. Here, on a summer's evening, the fragrance builds up to almost suffocating delight.

Between dawn and sunrise, nimble fingers pluck the corollas and truss them in loose sheaves for immediate transportation to the perfume factories of Grasse.

The delicate blossoms must be handled with great care. The fragrance is extracted by cold enfleurage which is a process reserved especially for Jasmine and Tubrose blossoms.

The blossoms are carefully embedded in layers of fat which has been spread on wooden frames. The frames are then covered with glass for further protection of the blossoms. The blossoms are turned every few days for at least three months as their fragrance is absorbed by the fat. Then the fat is removed from the frame and the essence is drawn off.

This is an extremely delicate task as well as time consuming. For that reason the price is high--from $5,000.00 to $20,000.00 per kilogram, which amounts to only about two pounds in our country. Expensive? Yes, but also heavenly!

There are over 100 species of Jasmine, so the assortment of oils is varied. This fragrance is extracted by cold enfleurage only, no matter where the extraction takes place.

Some chemists have created a "Jasmine-type" fragrance from benzyl-acetate plus a bit of civet which can be used in cosmetics and soaps. It is much less expensive than the natural oils of Jasmine, but it is certainly not to be compared with the scent of natural Jasmine.

During my research on this lovely flower, I found the Jasmine blossom passes through seven distinct stages of development from the time the tiny tight bud starts to open until it becomes a flat open flower. These stages compare with the seven stages of development we pass through in an entire lifetime. After many years, I was finally able to collect a supply of the Jasmine oil made from each of the seven stages. I then blended the seven oils together, in the proportions which pleased me, and named the resulting perfume "Jasmet," meaning the metamorphosis of the Jasmine blossom.

This is a light but extremely lasting fragrance. Those who have lived in the tropics, or ever known the joy of having a bit of Jasmine bloom in the evening of their life, are thrilled with this perfume--the memories are always pleasant.

Because the Jasmine is a "night-bloomer" its fragrance is controlled by the rays of the moon--rather than the sun--which makes it an exceptionally gay and happy scent, lifting the spirits and emotions of all to great heights! No one can possibly be depressed when wearing this lovely perfume.

There are some who firmly believe there is a similar note of fragrance to be found in the lilac, lily of the valley, even in Ylang Ylang. With this I simply cannot agree. It is most important to remember, however, that the Jasmine will blend with any other fragrance to advantage. It is most compatible.

Each time I inhale its lovely fragrance, I react to the words of an ancient Hindu poem, "From timid Jasmine buds, that keep/ their odeurs to themselves all day,/ But when the sunlight dies away,/ Let their delicious secret out."

We are told the Tahitian women bind the fresh blossoms of

the Jasmine around their necks and waists when they are being courted. Could there possibly be a sweeter scent to fall in love with? My answer is "No."

Lavender is another very old but still very popular scent. It is made from a plant which originally grew on the upper plateaus of eastern Spain and northern Africa.

We find, as a raw material in perfume making, the oil of lavender is considered one of the most important, especially in preparations to be used by men.

Sailors brought the lavender plant to England and later on found its way to the United States.

It became so popular in England that many believe to this day the plant is a native of that country. It isn't, but I do believe it is appreciated more by its adopted country! The English adore its fragrance.

For many years men came to me asking me to prepare a perfume just for men. You see, men who live abroad would never think of starting their day without perfume.

I recall one charming gentleman who told me that after living abroad for many years he had so thoroughly adopted their way of living that he would no more think of leaving for his place of business without applying the fragrance of his choice than he would think of starting out from his home without his trousers!

American men went through a period of thinking which caused

them to believe only sissies wore perfume! Which, or course, was a horrible thought! However, I find they are now slowly overcoming this morbid thought. (For which I am most thankful!)

As I say, from time to time men would ask me to design a perfume which would please them and still not cause them to smell like women. For several years, I passed this off lightly until one day in our office I realized I was being very thoughtless in this respect.

A patient had stopped in for a chat with my husband. As I came into the office with a case report, he asked, "Well, what's new in the perfume world, Mary? Have you ever designed that perfume for men I asked about a couple years ago?"

I had to admit, or course, I had not taken the time to actually work on it. He immediately became very serious and urged me to give it some thought.

"All right," I said, "I will make a bargain with you. Now, you are a most successful businessman. If you were to select a perfume for yourself, just what type of fragrance would you prefer?"

Instantly he replied, "If you will prepare a perfume which is 50% Havana tobacco and 50% Imported Scotch, I will buy it by the gallon!"

My first desire was to laugh--I thought it was funny--but, thank goodness, I saw the look on my husband's face and stopped short to think what this patient meant--before laughing.

You see, he was a heart patient, we had taken away his smokes and his drinks so he felt he no longer smelled like a man. The very things which had given him his courage and his strength had been taken from him. Now he felt lost and depressed. That was my cue. I promised him I would start to work at once on a perfume for men. I did that very night.

I began with a study of all the kings down through history and found, to my amazement, their favorite scents were lavender from England, heather from Scotland, and bergamot from Italy.

After many hours of research and experiments, I finally was able to blend these fragrances into a perfume which men have adored ever since. I have named it "Kings Choice."

Men wear it happily for it is truly a man's fragrance. Some women are very fond of it, too. I tuck tiny tissues soaked in "King's Choice" among my linens and our guests tell us we have the best smelling beds in all the world!

The perfume extracted from the lotus bud is heavenly. I was pleased to find this was not merely my personal opinion when I read a very old Arabian proverb: "The Lotus has its roots in the/ mud of the Nile, and its perfume/ at the throne of God."

Very little can be found regarding the lotus fragrance. It is not used in perfumes commercially sold today.

Yet the ancient Egyptians found it to be invaluable. It was always a part of the Royal Oils used for state and sacred occasions.

Because of its great emotional value, I have used it when blending my "Cleo's Kiss" perfume. This was one of the very first blends I ever made. It came to me as I studied the history of the ancient Egyptians.

While studying these interesting people in school, I had simply assumed they were a very brave, courageous and self-confident race. Then, when I began to study the fragrance of ancient Egypt, many years later, I learned this was not true.

The Priests knew of this shortcoming so, when preparing perfume for their warriors to wear into battle, they selected for that perfume the herbs and plants the fragrance of which, when inhaled by humans, gave them a feeling of courage, strength and self-assurance. The ingredients of this perfume were twenty-seven in number. Some were made from the bark, some of the leaves and some from the roots of the plants then available.

As I experimented, I found this to be true! The complete reaction was not present unless the lotus was included in the blend. The lotus, you see, is for strength--strength in the future yet to come!

Latin: Magnolia Grandiflora. French: Magnolia. German: Magnoliabüthen.

The Magnolia (Magnolia Grandiflora) is a native of south and central North America.

The flowers are very large, white in color, and look as

if they were made of wax.

In California, we had a very large and very old Magnolia tree which was much taller than our two story home. We always watched the buds grow fatter day by day. When they finally reached the stage of being "candle like" (a native term), we would take the long handled clippers (used in cutting off branches of orange blossoms, etc.) and carefully cut the branch just two or three leaves below the bud. This we would place in a shallow copper bowl. The place of honor for this beloved bud was always on a table at the foot of the winding staircase. There it opened in the night.

By morning, when we awakened, our entire home would be filled with this most heavenly fragrance! The cool night air wafted it into each and every room.

I find European perfumers invariably produce a perfume called "magnolia" by combining the true magnolia with other odors. This I also tried.

My blend, which I call "Touch of Spring," is made from the Jasmine bud, plus the yellow Jonquille of Holland, plus our own southern magnolia. It is fresh and clear as the first breath of Spring. Yet, when people sniff it for the first time, they will say, "Oh, this has a citrus fragrance, doesn't it?" It certainly does not. However, the combination of the magnolia and the yellow Jonquille plays tricks on us making us think of the sharp fragrance found in a citrus grove.

The leaves and the blossoms both give us a most delightful fragrance. Many times the scent found in the blossoms may remind one of a citrus grove--the same tang is there--but the fragrance remains forever in the memory of those who have lived in the south where this is a favorite tree.

When blending, the magnolia gives a certain softness, a tone that is both calming and exhilarating at the very same time.

It is greatly complimented by the Jasmine bud, given a more expensive effect by the addition of a deep rose, and given an expression of sheer luxury by the addition of only a wee dash of Tuberose.

Many times I have found when a blend is too ethereal, only a few drops (per pound) of the fragrance extracted from the leaves of the magnolia will bring it right down to earth. Thus, it gives the feeling of warmth and "belonging" that is most needed in our life today.

To my knowledge, there is only one fruit in all the world that yields two different spices plus a perfume. This miracle of nature is known as the fruit of Myristica Fragrance or,--the Nutmeg Tree.

The spices are called nutmeg and mace while the pefume is seldom mentioned today, but it's most valuable fragrance is well known and highly prized by all good perfumers.

We find this most valuable tree grows on the mountain

slopes of the West Indian Island of Grenada. Like the usual tropical tree, it is an evergreen.

When the fruit of the nutmeg tree ripens it becomes a very bright golden-yellow color--which is most striking--against the fresh smooth leaves as a background. When this beautiful fruit is ripe, the outer covering splits into two equal halves displaying the mace which is a curious network of bright red which covers the shell of the nut itself. The botanical name for mace is "Aril."

The mace is picked off and allowed to dry in the sun. As it dries, it turns a delightful shade of yellow. Amazing as it may seem, at this point the mace is worth three times as much as the kernel which is known as nutmeg.

When the Aril, or mace, is taken off the nut, or kernel, one finds the nut itself which must also be dried very carefully, before it can be cracked. When cracked, within this nut, is found the whitish solid seed which is penetrated by numerous delicate partitions of a reddish color very finely constructed.

The seeds are laid out on trays to dry in the sun. The kernel shrinks sufficiently as it dries that it rattles if the shell is gently shaken. It is then the shell can be broken and the nut or seed extracted. Women sit on very low benches, when the nutmegs are poured on to the concrete floors, and give each one a resounding tap. The shell flies in all directions exposing the valuable kernel. The seed is then put in bags and made ready for export to all the far corners of the world.

It is interesting to know a good tree will yield 1,500 to 2,000 nuts a year. The trees are planted about forty feet apart and should yield a harvest of about a thousand pounds of nuts per acre. It is said for the year ending June, 1958, that Grenada exported more than 523,000 pounds of nutmeg and almost 176,000 pounds of mace.

The essential oil which is expressed from the kernels or nutmeg butter seems to be identical with the oil taken from the mace itself.

Only a few drops of this precious perfume taken from the mace or the nutmeg often makes the difference noted in a most expensive and very rare blend of perfume oils. It lends a note of true quality, depth and body which can be found in no other fragrance. Yet it is never mentioned on the perfume counters. It is kept as one of the deep dark secrets of the perfumer.

Few realize the tropical orchid has given us one of the oldest and most delightful of all perfumes from ancient times down to the present day.

The true name for the tropical orchid is "Vanilla Planifola." First comes the breathtaking bloom--the aristocrat of all flowers--then comes the "bean" or seed pod which is the source of all pure vanilla extract.

From this vanilla bean is extracted, by pressure method, "vanillin," one of the most ancient of all perfumes. It has a soft, sweet and very pleasing fragrance.

"Vanillin" is one of the ancient fragrances I have blended together with fifty-eight others to make my "Persian Nights" perfume. I recall, after one of my lectures, one lady came to me and said, "Do smell my arm where I applied "Persian Nights. It smells exactly like vanilla ice cream."

Her excitement was well founded, and she was exactly right. Her particular body chemistry immediately brought out the fragrance of the vanillin or orchid.

Only the seed from the White Orchid gives us a fragrance which is celestial in its attitude, not too sweet, extremely light and perfect for brides!

When asked to appear on television as guest of Dorothy Fuldheim, News Analyst for Scripps-Howard Broadcasting Company, Channel 5 of Cleveland, I decided it would be great fun to make a television first by making perfume on the program for the very first time. This was so exciting! The program stirred such interest in the World of Fragrance I received letters, calls and comments for months. You see, I started by crushing the seed pod of the wild white orchid sent to me by air mail from the South Sea Islands and by the end of the program shared with Dorothy Fuldheim my "White Orchid" perfume. She inhaled deeply and spontaneously exclaimed, "Oh, this is heavenly!" One of the great thrills of my life!

I shall explain this in detail in a later chapter entitled

"Methods of Extraction." I hope you, too, will find it of interest.

I found the history of the orchid, "The Aristocrat of the Woodlands" to be most enlightening.

This flower comes from a very large and most complex family. Botanists have discovered and named over 16,000 different species of orchids.

Various members of the family thrive in most places on earth. Some are perched atop giant trees, sit prettily on plateaus and cliffs, or grow from lofty vertical rock faces. Some are found in the beds of streams, or on the rich, leaf-mold of the forests. While others are found among the clouds on the snowy heights of the Andes or even higher up in the Himalayas, thousands of feet above sea level. Other members of the family make their home in the warm valleys of Mexico, Nepal and Java, as well as the icy forests of northern Canada.

I had always supposed the orchid was a tropical flower, requiring a warm, damp climate. I was greatly surprised to learn they are not choosy about where they live. All they actually require is fresh air, water, food and sunlight in the right amounts at the right time.

They will grow as a houseplant in a kitchen window, in the coconut husk of a native, in a ship cabin, in a greenhouse or in a garden that is not too dry. It would seem they will grow anywhere just as long as they are wanted or really needed.

This flower is noted for its grace and beauty. Its color ranges through every known shade and every conceivable hue, all the delightful colors of the rainbow. We are most familiar with the lavender, the white, and the purple. But do you realize there are orchids which are empire green, pink, chocolate brown, mustard yellow, chartreuse, apricot yellow and bright red?

The fragrance of all, when the flower first opens, is heavenly, while some have a most powerful and lasting perfume. Their beauty is frequently found at weddings, anniversary parties, and fine balls, causing us to think of them in connection with romance.

Orange Blossoms--Latin: Flores Aurantii. French: Fleurs d'Oranges. German: Orangenbluthen.

The flowers of the bitter orange tree (Citrus Vulgaris), as those of the sweet orange tree (Citrus Aurantium) contain most fragrant essential oils. However, the fragrance differs in flavor and value according to their source, and certainly by the method of preparation.

The leaves of both trees also contain a most unusual oil often used in the past of perfumery.

The odor of orange blossoms always reminds me of brides! It recalls to my mind the delightful moonlight walks through the citrus groves during the years of my life in California. We had several different kinds of orange trees in our grove, five if I remember correctly. I trained myself to recognize the scent of each different blossom--even with my eyes closed. At quite an early age, this became a most delightful game with me. How thankful I am today for that early training of my olfactory nerve.

It is a well known fact that the blossoms of the orange tree should be gathered late in the morning when the weather is warm and dry. Usually, where there are citrus groves, there is frequently a late morning fog which is cool and damp. This must burn off before the orange blossoms are picked because the dry blossoms yield a much greater percentage of oil. Perfumers know the oil of orange blossoms is obtained from the blossoms, but the famous oil of petitgrain can only be distilled from the leaves or tiny twigs of the tree; the oil of orange is always made from the rind of the fruit. These are busy trees!

As a child I remember squeezing the rind removed from an orange and being amazed at the amount of oil which would suddenly appear. One day my father took a bit of the rind from my hand and with a match set it burning. As the oil burned, a wonderful fragrance filled the air. How excited I became over this experiment!

In the years which have followed, I have often captured this precious oil, added it to wax and burned it as a means of perfuming the air.

The one odor which I have found to be similar to a degree is bergamot. This, of course, is truly a citrus as it is a tiny sour orange which grows in the very tip of Italy. It is not related to our flower-garden-bergamot at all. The fragrance of both lime and lemon blossoms are faintly similar but not nearly as deep or comforting as the sweet orange blossom oil.

Patchouly or Patchouli--Latin: Pogostemon Patchouly. French: Patchouly. German: Patschulikraut.

Patchouli is an herb. It was first found in India, where it has grown for centuries by the Ganges, in the East Indies and China. As it grows, it greatly resembles our sage plant.

We are told the ancients gathered the leaves of this herb, placed them in net bags which they hung in each of the corners of their rooms to drive out insects, mildew and evil spirits!

It was my good fortune to come upon a supply of the patchouli leaves which I found to be most interesting. However, the fragrance obtained from those innocent little leaves was so very potent, it took seven years of hard work before I managed to make it into a perfume which could be worn to advantage in this day and age.

I shall never forget the year I brought it out as my new Christmas fragrance. The reaction was so amusing! Some found it overpowering, some spent hours trying to recall just what it was this new scent reminded them of in their past, and others adored it immediately. The most outstanding reactions were this: Every single person who sniffed patchouli found it to be most interesting. It would not be ignored!

It has long been used by the countries of its origin to perfume Cashmere shawls, India ink, as well as many other things offered for export. Perhaps it was grandmother's shawl--so comforting in your childhood--which first brought this unusual fragrance into your life. Who knows?

The shawls of India have been a favorite luxury in Europe for generations. The French found they were quite able to imitate the shawls but they did not sell. The French shawls lacked the fragrance which make them genuine and clung to the shawls forever. It was many years later before the French manufactuers discovered the well guarded secret. It was then they imported some patchouli plants from India.

After considerable effort, they were finally able to make them grow. But the art of knowing exactly when to collect the tiny leaves, the process of drying them, the art of extracting the fragrance from the leaves required many years of experience. When, at last, their goal was reached, the French were able to make, and offer for sale, shawls so closely resembling those famous shawls of India it took a real expert to detect the imitation.

The stems and leaves of this precious plant must be gathered for drying and extraction before it reaches the flowering stage. This is the secret!

In India the patchouli is known as "Puchaput." They have used it for centuries in sacraficial incense--both in sticks and pastils--for religious purposes.

This unassuming little plant brings us a very heady, pungent, almost overpowering fragrance.

My "Patchouli" blend has been beautifully received! Those who wear it most frequently tell me it gives them a great feeling of security in these trying times.

One dear friend has a lovely vase from India on her shelf of a room divider wherein she has tucked a bit of cotton saturated with my "Patchouli." The moment anyone enters her home, they are greeted with this delightfully different fragrance. There follows, unconsciously, a feeling which tells them all is well. The welcome the scent extends is relaxed security for all who enter.

The most wonderful rose perfume in all the world comes to us from Bulgaria. No other country can hope to compete with this delightful fragrance.

The rose industry, which was brought to Bulgaria in the 17th century by the conquering Turks has flourished ever since. The Turks, evidently, learned the art of distilling rose oil in Persia. Because of the climate, however, the soil and the abundance of peasant labor, the raising of roses is much easier in Bulgaria than in Persia.

The Bulgarian rose growers always pray for a damp season, not only because they know the rain intensifies the fragrance, but, also, because too much sun makes the roses grow faster than they can be handled.

In the wee hours of the early morning, long before dawn, it is a common sight to behold the Bulgarian peasant women aglow with color in their native costumes busy in the rose fields picking the blooms and piling them in the horse-drawn carts which take them to the perfumeries.

It is said the maximum yield is around two o'clock in the morning. You see, they must be picked, loaded in the carts, and taken to the distillery before sun up or the roses are ruined.

It takes about 4,000 pounds of Bulgarian rose petals to produce one pound of rose oil or "attar of roses."

The medicinal value of the rose is great. History mentions thirty-two remedies made from the petals or leaves of roses. At one time, we are told, rose petals were used to flavor wines, salads and conserves.

The words "Attar" and "Otto" are synonymous in meaning "Essence" when distilled. However, when other extraction methods are used the oil is correctly called "Absolute."

It is interesting to know that no two rose oils bring us exactly the same fragrance. Each rose gives us its own individual and characteristic perfume. I find they also vary greatly in color, weight, strength and enzyme activity.

Exactly as the ancient Egyptians extracted their perfumes, do we process the rose. The petals are placed in huge copper kettles, water is added, then the containers are made air-tight while underneath are started wood fires. As soon as the distillation begins, the heat breaks open all the tiny cells releasing the perfume oil which soon floats on top as the water sinks to the bottom of the still.

In the beginning, perfume was distilled from the rose petals placed in highly decorated, copper, open fire stills. Today the method is much the same, only modernized. The stills are much larger now, and each one holds almost a ton of the precious rose petals per run.

It is believed it is the phenylethly alcohol <u>naturally</u> found in rose petals which gives the fragrance its soft mellow depth. During the last few years chemists have done extensive work in trying to create this substance synthetically. At long last, they have been quite successful in preparing phenylethyl alcohol from coal tar derivatives. But there is no comparison!

I had a most interesting experience with this new discovery. One of my importers brought three tiny vials of rose oil for my inspection. I sniffed one and found it most pleasing. I then sniffed the second vial and found that to be even lovlier! Then I sniffed the third and began to sneeze with terrific violence. It frightened the importer almost out of his wits! But not I, for I had had similar experiences before.

I told him, after I recovered my breath, I would like a supply of the second vial but never to bring me synthetic oil again as long as he lived. He was amazed and still more frightened that I recognized a synthetic so quickly. He asked, "How did you know? It must be magic!"

I have never explained it before, but I feel that God has given me this "magic" as a protection. You see, I am terribly allergic to all coal tar products, drugs and synthetics. Often my importers, thinking they can fool me, will slip in a synthetic but, thanks to my super-sensitive nose, I have no trouble sorting the real from the unreal.

Making Attar of the Roses through my "Enzyme Brew Method" has been a real joy. My husband's hobby is growing roses--how fortunate for me! Bless his heart, he is most generous with those precious buds, for he realizes I have found a way to make them live forever.

We watch the buds and when we see they are about to open, I take the flashlight at 1:30 in the morning, hurry out to gather the buds, and set the brew immediately!

How fascinating it is to watch the perfume form. The seven drops (or seven times seven depending on how many buds are set to brew in one sterilized bottle) soon become an ounce. That ounce becomes four ounces in the course of a few months.

Many believe the Peace Rose has little, if any, fragrance. Some of the most delightful Attar of Roses I have made were made from the buds of Peace Roses grown in our garden. The color is retained by the enzyme reactions and the fragrance enhances with age! Truly delightful!

My husband grows a beautiful Persian Rose so deep both in color and fragrance that it fairly thrills you to come in contact with it! The fragrance brew from this particular rose is, I believe, the most exciting of all.

In comparison, the Peace Rose is an aristocrat, aloof, and in a class all by itself. While the Persian Rose is filled with deep warmth, unspoken desire, and as compelling as a magnet.

When men inhale the fragrance of the Persian Rose they say, "<u>This</u> smells good enough to drink!"

In years long gone by, we are told, they drank the Attar of Roses if they could afford the luxury. It gave them a nice glow and how sweet their breath must have been!

Another well known flower oil is rosemary. It seems it was originally called "Ros Maris" which meant "sea dew."

This fragrance was first made famous during Queen Elizabeth I's time. It was then that Shakespear wrote, "There's Rosemary/ That's for remembrance."

It once again came into vogue during the 19th century and all the fine ladies of that time went wildly mad over rosemary, even here in the United States.

The scent today is still used in very expensive soaps.

Perfume is made from both the leaves and the flowers of this plant. It has remained famous all these many years simply because of its refreshing--yet stimulating--fragrance.

The dried, crushed leaves of the Rosemary, gently sprinkled over a hot broiled steak is the dream come true of any gourmet. How beautiful the fragrance enhances the taste.

In a book printed over 500 years ago, I found the following suggestion, "Take the flowers of rosemary and/ make powder thereof. Bind the/ powders to thy right arm with a/ linen cloth, and it shall make/ thee light and merry." Also, "Make a box of rosemary wood,/ and smell it, and it shall make/ thee keep thy youth."

It would seem rosemary is not only an herb of rememberance but also a plant of great promise as well!

At the mere mention of the name "Sassafras" most of us are reminded of the small bundles of this wood, derived from the root of the American Sassafras Officinalis, which used to appear quite frequently in stores, especially in the spring.

From this wood a tea was made which was given to many of us during our childhood. It was served both hot and cooled, with ice as a luxury. All this happened during the days before soft drinks became so plentiful and popular, and also before the day of the ice cube!

I especially enjoyed sassafras tea when a little honey was added while the tea was hot, then allowed to cool. Yes, this was a favorite treat during my childhood. The memory, which often returns to me, does not pertain to the delicious taste alone; it was the fragrant aroma which delighted me most of all. Imagine my surprise to find sassafras, this old friend of mine, also lived anew in the World of Fragrance!

We are told the Europeans made sawdust of this root, then mixed it carefully with pine sawdust, moistened it with fennel water and allowed it to dry. The fragrance, extracted from this sawdust, so prepared, is used in some of the most expensive soaps. This extraction, or oil, is called "safrol."

In later years, the Japanese discovered there was a trace of "safrol"--very similar in fragrance--to be found in the crude oil of the Japanese Camphor. The latter was much more plentiful, much cheaper in price, and very soon replaced the natural oil of the sassafras on the open market.

The beautiful Tuberose plant is known to be a bulbous amaryllidaceous herb cultivated mostly in the south of France for its spike of white lily-like flowers.

The buds of the Tuberose should be gathered early in the morning when the buds are just about to open and they are still slightly damp with dew. Many extract the oil by means of enfleurage but, again, I find my "Enzyme Brew Method" to be far more satisfactory.

It is interesting to know that the Tuberose is the fragrance sold most often by the perfumers as Gardenia. The oil extracted from the Gardenia does not have the same fragrance as the Gardenia blossom itself. However, the oil extracted from the Tuberose is an ideal way in which to capture the true fragrance of

the Gardenia in an often make-believe world. The World of
Fragrance is filled with mystery...this is one of them!

Tuberose alone is almost too overpowering--sometimes to the
extent of being depressing for some. Therefore, it is wise to
blend it with Jasmine, Cassie, Orange Blossom or even a bit of
Magnolia for more perfect results.

We think of thyme today as a seasoning for our food. In
ancient times it was used extremely much as a perfume.

The ancient Greeks used it as an incense in their temples.
There are many legends which give it a most attractive history.

Shakespeare wrote much about this scent. He wrote in "A
Midsummer Night's Dream" a lovely passage which goes, "I know
a bank where the wild thyme grows..."

Evidently, he had also read the delightful legends of the
past which lead us to believe the fragrance of thyme on the air
on a midsummer night could lull one into a wonderful dream where
amazing sights could be seen. Such as fairy kings, queens and
a procession of elves and sprites could be seen, making cares
flee before entering a world of fancy.

It is also said that thyme was one of the three plants which
made the bed in the manger at Bethany for the Virgin Mary and
the Christ Child.

The Aromatic Vinegars were really more refreshing, according to the ancients, than our toilet waters, colognes, and perfumes made of alcohol.

One of the ancient recipes which has come down through the ages is called "Toilet of Flora." It reads, "One handful of rosemary, wormwood, lavender and mint to be placed into a jar with a gallon of strong vinegar, kept near a fire for four days, strained, and an ounce of powdered camphor added and then bottled for use."

You can easily imagine the variations resulting from this recipe, depending entirely upon the size of the maker's hand.

Most of the violets which are grown for perfume are raised in southern France. They grow in shady places; in Grasse they are usually planted at the foot of the trees in the groves.

The harvest usually runs from February to April. The blossoms are gathered two or three times weekly and are always picked just after sunrise while they are freshly sprinkled with dew.

Either extraction or maceration is used to obtain the perfume which is a most delicate and highly expensive oil. Why? Because many reports show us an entire acre of the lovliest vioets will yield only a few precious drops of the perfume oil!

Some perfumers also use the oil extracted from the violet leaves. In my experiments, I have found just a tiny bit of the oil of the leaves added to the perfume extracted from the blossoms gives my perfume of violets a much more woodsy fragrance, adding greatly to the depth of the scent.

Often a few drops of violet oil, when added to a blend of many flowers, gives a desirous effect. The note of interest it projects is subtle, reminding one of the breeze wafted from the woods in early spring.

Many perfumers employ only the Parma Violet or the Victoria Violet, but I have found the Sweet English Violet to be warmer, much more dream provoking.

Ylang Ylang (ee-lang ee-lang)--This is a delightful greenish-yellow, or chartreuse, blossom with a lovely red center. The flower droops, but is called in the orient "Flower of Flowers." The Polynesian women wear these blossoms in gracious garlands at all celebrations. It is a very strong flower, so strong, in fact, that it can be subjected to steam distillation for the removal of its perfume. This fragrance is excellent for oriental blends, gives any perfume of that type a delightful lift, and still keeps them mellow and soft.

This is considered the sacred perfume of the orient. For hundreds of years its gracious fragrance was used only to anoint the robes of the Mandarin the day the robes were dedicated. After the ceremony, the robes, we are told, were placed in chests made of fragrant wood where they rested until an important ceremonial demanded their use.

One of our doctors invited us to his home for the evening to honor the return of his son who had been stationed in the orient during World War II.

During the evening the son brought forth many gifts--great treasures--which he had purchased in the far east. Carvings, paintings, lamps of great beauty. Then he brought out a large bundle wrapped in heavy parchment. As he opened it, there were countless "Ooh's" and "Aah's" gasped. There before us was the most beautiful of garments. A Mandarin's robe of gorgeous satin embroidered in threads of gold, silver and many rich colors. It was a sight none of us will ever forget. The thing that impressed me most of all was the perfume which immediately filled the room. The scent of ylang ylang laid heavily in the air.

The robe had been dedicated at least 350 years before, but the glorious fragrance was as fresh as the day the robe was anointed.

The more delicate women present found it most distasteful, and at first they complained bitterly. Then, as the evening wore on, I noticed those who had complained the most saying, "That fragrance...as one becomes accustomed to it...it seems quite pleasant...but, oh, so strong!"

This gave me the idea that, perhaps, a place could be found in our world today for the precious ylang ylang. The seed was planted!

At one time, a year or so later, I was able to obtain a tiny bit of the raw ylang ylang with which to further my experiments in this field. I found it to be most powerful, almost too potent, and far too overpowering for the ladies of my acquaintance. It took me over seven years to finally find the fragrances with which I could blend ylang ylang, taming it sufficiently for modern enjoyment. What joy that discovery brought to me!

Can you imagine what the little helpers were? They were the fragrances extracted by my "Enzyme Brew Method" from the tiny flowers which preceed the holiday berries, the holly, the mistletoe and the bayberry. I gave this to my friend as a Christmas gift and called it "Merry Christmas." They have loved it for many years because it is filled with interest and joy. It can be worn at any time during the year--to great advantage.

Down through the ages there have been found various sources of perfume. Of course, the only true perfume is that extracted from the living things of this world. Let us now delve a bit deeper into the exact sources of extraction.

We find the fragrance exists in the minute traces of essential oil which is found in the petals of the rose and the lavender, and occasionally in the glucoside which is decomposed in the presence of an enzyme or ferment as in jasmine and the tuberose.

It is the <u>flowers</u> of the cassia, carnation, clove, myacinth, even the heliotrope, mimosa, jasmine, jonquille, orange blossom, rose, violet, and the ylang ylang of China which give us their lovely perfume.

From the <u>flowers and leaves</u> of lavender, rosemary, peppermint and violet is extracted an earthy fragrance.

It is the <u>leaves and stems</u> of geranium, verbena, and cinnamon which give us their perfume.

An even stronger, more potent perfume is extracted from the <u>bark</u> of the camella, cinnamon and cassia.

It is from the <u>wood</u> of the cedar, linaloe and sandalwood that the fragrance is extracted.

While the perfume of the angelica and sassafras is found in the <u>roots</u>.

The <u>rhizomes</u> of the ginger, orris and calamus are famous for their most delightful and treasured fragrances.

The <u>fruits</u> of the bergamot, lemon, lime and orange give forth their perfume.

While the <u>seeds</u> of the bitter almonds, anise (both kinds), fennel, nutmeg and white orchid are famous sources of perfume.

<u>Gum or oleo-resinous exudations</u> from the labdanum, myrrh, olibanum, Peru balsam, storax and tolu have come down through the ages as a most valuable source of perfumes and perfume bases.

The orient will always be famous for its own particular type of exotic fragrance. But it is southern France which will always be known as the Paradise of Perfume!

Bulgaria has justly gained a special distinction for the extremely fine quality of the Attar of Roses distilled from its famous rose gardens.

From the delightful villages of Mitcham and Hitchin in England we learn of a fame carried down through the centuries pertaining to their secret of extracting the oil of lavender and the oil of peppermint.

Cannes is most famous for its rose, acacia, jasmine and neroli oils.

Nimes shines forth with glory in the field of thyme, rosemary and lavender--so different from the lavender of England.

Nice holds only one claim to fame...its famous essence of violets.

Grasse is known the world over for its perfume "par excellence." Many have described it as the "heart of sweetness of the modern world."

I have always marveled at the ability of those French gardeners. They are so proficient in the art of gardening that they are successful in producing flowers twelve months a year! Have you ever stopped to wonder why some people you know can make any plant grow, and others you know have no luck at all? I have found, when you talk quietly, calmly, with anyone who is noted for having a "green thumb," you will find that they actually talk to their plants, encouraging them to grow and do their best!

Call it "witchcraft," or whatever you wish, but I call it "The Power of Suggestive Growing," and I know it really works.

This may seem strange to some, but it is the most interesting experiment I have ever tried. My results have proven to me, without a doubt, that plants, and all living things, are not without a soul, thoughts and gratitude.

Upon our semi-retirement, my husband, who had never gardened, became extremely interested in growing beautiful flowers, especially roses, shrubs, bushes and even trees. He is 100% Holland Dutch with a very strong ability for projecting his thoughts, his will or his ideas.

I was amazed one day to overhear him say to a flowering shrub, which after three years of the most loving and skillful care, had not produced one single blossom, "I will give you just one more year! I am telling you this, if you do not bloom this year, I shall tear you out by the roots!" He meant every word it too!

Evidently the plant recognized this as an ultimatum, for that year it bloomed as I have never seen any shrub bloom in all my life! It definitely knew who was boss!

Chapter 16

Fixatives

There are only four animal fixatives of real importance to the perfumer: musk, from the musk deer; ambergris, from the sperm whale; castorium, from the beaver; and civet, from the civet cat.

Many have asked, "What does a fixative do, make perfume go further?"

My reply is "No!" A fixative makes a perfume last longer and do more as it goes.

When an animal secretion is added to perfume it not only "fixes" or "sets" the fragrance which greatly increases its lasting qualities but, in my mind, it also gives the perfume a "zip" which quickens the pulse, and may even take your breath for a moment with its unusual and interesting approach.

Many times I have blended a heavenly floral perfume. As my friends sniffed it they would remark, "How nice! Or, interesting, isn't it?" But never a word about taking it home with them.

Then I have added a drop or two of the fixative suitable for that type of fragrance and when next they sniffed it they would exclaim, "Wow! Terrific! I simply must have some of that!"

Because I have seen this happen time and again, I am of the opinion it is the fixative, when properly selected and blended, which creates the desire to have and to hold.

Selecting the proper fixative for each blend of perfume, as it is originated, requires endless time, the greatest of skill (acquired only as a result of years of experimenting) and extreme sensitivity. The true magic touch lies in the fixative selected. It is, indeed, the animal scent which makes a perfume ordinary or simply out of this world. Still, it must be so cleverly concealed that no one would ever suspect its presence.

Let us now delve deeper into the source of these interesting "animal perfumes."

Musk--We find perfumes are divided into two great classes. Those made from animal fats and secretions, and those which are derived from vegetable matter such as herbs, plants and flowers.

The very oldest, and by far the most famous, of all the animal varieties is musk, which is a fixative obtained from the musk deer.

Musk is a dry secretion extracted from the preputial follicle, or musk bag, which lies under the skin of the abdomen of the male musk deer.

The animal is a native of the higher plateaus of the Himalyan Mountains. The most valuable and popular come from Tibet.

The ancient Monks of Tibet were the first to realize when God placed fragrance in flowers to attract insects to carry on pollination, He also placed a little perfume sac within the safekeeping of some animals. Most of these animals live in rugged climates and are easy prey to their enemies. Thus, it is necessary for them to mate more than once a year. In this respect, the perfume is of great assistance. In most cases, it becomes apparent only at the mating season.

As a result of their discovery of the musk pods, the Monks of Tibet have made fortunes down through the ages, with which to carry on their charity work, all from the sale of musk to the perfume industry.

We are told the musk deer is a hardy, solitary creature, quite timid and never found in herds. Only rarely are these animals found in pairs. Evidently, the male prefers to have his mate remain in seclusion. During the summer months, he is seldom found at an altitude lower than 8,000 feet. Dense thickets of juniper, birch or rhododendron offer perfect protection for his home. The hunters have found various ways of tracking down the musk deer. I believe this method to be the most interesting of all.

It seems the male is extremely fond of the melodious sound of a flute. Therefore, it is most advantageous for a hunter to be accompanied by a flute! Should he not be this fortunate, it is then necessary for him to hire out such a musician to conceal himself in the thicket and play the flute.

Being an extremely curious animal, the musk deer will often throw caution to the wind, approaching nearer and nearer to the source of the melodious tune. Needless to say, this love of music is his downfall.

We are told the most prized animals are bucks from ten to twelve years of age. At one time, long ago, it was possible to realize two ounces of pure musk from one such animal. The young musk deer yields much less in amount. It is a pity to kill them since the yield is so very small per animal. The females have no perfume pods at all.

As soon as the animal is killed the gland containing the musk is immediately removed in its entirety. The ancient Monks of Tibet dried these glands in the sun as a means of preserving them until the arrival of the traders. In later years, to hurry the process, they were dried by the heat of small stoves, and some went so far as to immerse them in hot oil.

The very finest musk in the world is called "Tonquin Musk," and comes from Tibet. This, by the way, is by far the most costly, too.

There are many stories of dishonesty in the history of musk down through the ages, resulting the traders becoming very wary at the time of making their purchases. It was the Chinese who found a way of reassuring the traders.

They discovered the musk pod, which is actually a small sac, almost round in shape, hairy on one side and quite smooth on the other, contained a very small orifice in the middle. With a very small instrument, which they call a "sonde," they found they could enter this tiny opening, extract a bit of the musk grain, let the trader examine it to prove its high quality, replace the musk within the sac and one could never even tell it had been disturbed. The Chinese are extremely clever people. In this way they were able to guarantee each and every sale made to the traders. Consequently, they controlled the market!

The fresh musk pods certainly would never remind anyone of perfume. It is necessary to remove the outer covering of both hair and skin first, then soak the gland in water for quite a while before it comes soft and the odor becomes agreeable and pleasant.

Often through the ages it has been thought a new source of musk had been found. A secretion extracted from our own Florida alligators caused quite a flurry only a few years ago. However, in spite of constant search, the true musk from the little musk deer continues to retain first place in the perfume world.

Surprisingly only a few centigrams of musk will fill a very large room with its overpowering odor without showing any appreciable loss in weight whatever. When using musk in perfume, it must be remembered a little goes a long, long way!

An amazing thing is that musk is not only used in the perfume world; it is also well known for its great medicinal value. It is said that even today it is used in over fifty-six different medicines. It is used as a stimulant and also as a sedative. The dosage governs the reaction, in perfume and medicine.

Enchanting stories are told of the association of musk with the perfumes of ancient times. It is said, in the building of a certain famous mosque, large quantities of musk were mingled with the mortar so its fragrance shall remain forever.

Ambergris is another old and renowned "animal perfume." In the perfume world it is pronounced "amber-gree." It is known to be a waxy concentrate formed in the intestines of the sperm whale. Some have even gone so far as to add "a sick sperm whale," but I prefer thinking the whale, who cast forth the ambergris, was extremely healthy and wonderfully happy. As happy as his scented wax has made me as I have worked with it.

We find this fixative mentioned in history ever since the latter part of the 15th century. It has the most romantic history of animal fragrance in all the world!

Its very name is romantic. In Arabic it means "perfume," any perfume.

Ambergris is the most wonderful fixative ever discovered for perfume. It is the one ingredient which the chemists have never been able to make by artificial means, nor has there ever been a synthetic substitute found to replace ambergris to this date.

The great secret of ambergris is that it has very little, if any, scent of its own. The slight odor it does have is slightly fishy, very similar to linseed oil but wonderfully sweet. Therefore, it may be combined with the most delicate fragrance without changing the scent in the least. For that reason, the most expensive of all perfumes always, in years gone by, contained this fixative. The reason is that ambergris is the only known substance which will definitely "fix" a scent without altering its odor at all.

Throughout the ages, it has remained a mystery as to why ambergris should possess the strange property of fixing a perfume and causing the scent to linger on for a considerable time after exposure to the air. Of course, ambergris is a most mysterious substance. Even its exact composition and the cause of its origin are to this day complete mysteries.

It is known that ambergris is formed within the intestinal tract of the sperm whale, but it does not appear to be a natural condition, for it does not exist in more than one whale among literally thousands!

Because of this, some believe it is caused by a diseased condition suffered by the whale. Often ambergris has been found to contain portions of the horny beaks of giant squids, or cuttlefish, on which the whale feeds. For that reason, some believe it results from an irritation caused by the more digestible portions doing their job, but the indigestible parts of the squid accumulating in large masses. No on knows why some whales have great masses of ambergris while others have absolutely none, for they all feed on the same diet and live in exactly the same manner.

Even this origin does not make the ambergris the slimy fibrous material one would expect. It is fairly firm, often quite porous substance that is slightly greasy to the touch. It is usually of a grayish color, but it may also be black, white, brown, yellow and even a mottled combination of these colors. For this reason, it is most difficult to recognize.

Most of the ambergris obtained so far has been found cast up on the beaches or floating on the surface of the sea. The very thought of discovering this most valuable substance lends great excitement to the lives of the captains and crews of whaling vessels. They eagerly hope each whale they take will be the one to furnish a fortune in ambergris.

The weight of the ambergris found in whales seldom is more than a few ounces--a pound or two at the most--while the deposits found floating on the sea, or washed up on the beaches, often weigh 100 pounds or more. It is listed that during the fifty years from 1864 to 1914, 2,576 pounds of the precious ambergris were brought to New England by whale ships.

The sad thing is that many sailors have found a mass of ambergris washed up on the beaches, but have passed it by without thinking of it as anything with value.

I found one such story to be of great interest to me. Years ago, while several Hawaiin cowboys were bathing their ponies in the sea, they noticed some masses of what they thought was sponge floating in the water. They began to use them to wash and wipe down their ponies.

They soon realized the material was not sponge and began to wonder what in the world it really was!

They carried off some of it to a local merchant-trader for his opinion. Just imagine their amazement when they learned the substance was really ambergris.

They hurried back to the sea where they had found it in hopes they could capture the remainder of what they had left behind. In the meantime, much of it had been carried away by the tide, but even the small amount they had managed to retain was sold for enough to make them all independent for the rest of their lives!

There are countless stories throughout history telling of sailors who were pitifully poor until the day they found even just a handful of ambergris, making them extremely wealthy for the rest of their lives.

In ancient times the very hope of such a find kept many a sailor from being lonely and depressed. It added greatly to the glamour of sailing the seas.

Isn't it wonderful to realize how this substance has been surrounded with romance from the time it left the intestines of a whale until it reached the courts of kings?

It could only happen in the history of perfume.

Next in importance, as a fixative, after musk and ambergris

comes civet.

This we find is a thick-yellowish, fatty secretion of a strong musky odor. It is obtained from glands in the anal pouch, located near the sex organs, of the civet cats.

The civet is different from the rest of the cat family. It does not climb trees! It is as much at home in the water as on dry land. We are told it is short-legged, long-bodied, broad-faced, round-eared, bushy-tailed and an unfriendly looking cat! In fact, some describe it as quite "fierce." It is from ten inches to one foot high, and about three feet long.

Our first encounter with this cat in history dates back to the earliest part of the 16th century.

The thing I like about civet is that the animal does not have to die to give it to us. Not that I imagine he enjoys giving it to us but, at any rate, he can still continue to give it and go on living happily.

The Dutch made a business of raising the civet cats in cages which were too small for the animal to turn around. They fed them a diet of mostly milk and egg whites, which caused the deposit of civet to become white rather than the undesirable brownish-yellow which was its natural color.

The pouch, which contains the civet, is in two parts and located near the anus. Every few days an attendant would tease the cat (the more annoyed the cat becomes the more civet his little gland produces), then through the bars of the back of the cage he would remove the civet by inserting a small wooden spoon in the orifice of the pouch. It probably was not too comfortable for the cat, but we are told they all lived to be a ripe old age.

Before this method of extraction was discovered, the civet was found in small pellets scattered around the lair of the civet cat.

When the civet pots, containing the civet extracted by the Dutch, went to market they carried a written guarantee assuring the purchaser of its absolute purity.

Both the male and the female cats appear to have this endless supply of civet.

In ancient times the crude unrefined civet was carefully packed and shipped in the hollowed out horns of the male Zebu. A very suitable container when you really stop to think about it!

Of all the animal substances used in making perfume, civet has the most revolting odor.

However, when used in a well diluted solution, the disagreeable odor is lost, making civet an extremely valuable fixative.

This interesting fixative, as far as I am concerned, has one distinction: when only a very few drops of civet are added to an exotic blend, the results are positively tremendous. The perfume suddenly becomes as wild as a march hare! My "Gypsy Violin" is a living example.

For the blonde, delicate timid type, it holds no charm. In

fact, it frightens them. They do not know how to cope with it.

For the dark, mysterious exciting type, perfume containing civet opens new doors and presents endless dreams filled with thrills and excitement. To each her own--the World of Fragrance holds something for everyone.

Closely following civet in importance comes a very strong smelling, oily substance obtained from the sexual glands of a beaver. This is known as castoreum.

This is used as a fixative in the perfume world, and also as a stimulant in the medical world, which gives it a double value.

Castoreum is first mentioned in perfume history during the 17th century. It was most expensive then, but now the pelts are worth more than the castoreum!

There are two pear shaped membraneous sacs found in both the male and female beaver near the genitals, which contain a creamy secretion. During the life of the beaver this serves as a coating for the animal's fur and tail, making them waterproof. Without this protection, their coats would become water soaked and they would never manage to swim or float as they worked on building their dams.

When the sacs are removed, from the now dead beavers, the secretion is removed and dried either in the sun or smoke dried over burning wood.

As the secretion is removed from the sacs, the odor is

nauseating, but after it is dried and properly prepared for the perfumer, it is quite pleasant, giving off a sharp aromatic fragrance all its own.

It must be remembered that fixatives must be used with utmost discretion. The ideal fixative is of a retiring nature. Therefore, the type to be used must be selected with great care and the quantity added must be guarded with even greater care. Just one drop too much can easily ruin an entire batch of blended perfume.

There are fixatives which blend very well with heavy exotic fragrances, but which would destroy completely the more delicate fragrances. I have found it is much wiser to add only very small quantities at first and gradually, after considerable aging, increase the proportion until the desired effect is finally obtained.

Observing a fixative do its chosen work is a most fascinating thing to watch!

The ancient Egyptian and Greek perfume makers were not only limited in technical knowledge, but were also extremely limited in materials as well. You see, they knew nothing of animal fixatives. It was long after their first experiments that they found the Chinese had discovered the perfume which had been given to animals.

Therefore, at first, they were forced to find and employ <u>vegetable fixatives</u>. This must have been a real challenge.

Yet, how very adept they were!

They extracted their vegetable fixatives from plants containing resinous substances such as balsams, gums, and oleoresins.

It is amazing how they learned, simply through trial and error, the essential oils to be used as fixatives could be found in the oils bois de rose, clary sage, orris, patchouli, sandalwood, and vetiver.

Bois de rose and clary sage have a sweet flowery type of fragrance and can be used with perfume of a more delicate bouquet.

Orris and sandalwood give a heavier fixative and can be used with the violet type perfumes.

Patchouli and vetiver are pungent in odor and can only be used in oriental bouquets.

The ancients had a "tablet of guidance" which must have been a priceless guide for early perfume. Some of this early knowledge reads as follows, "Balsam of Peru is an excellent fixative for rose, heliotrope, magnolia and lilac perfumes.

Benzoin is a common fixative and has a vanilla like odor. It blends well with almost any type, such as violet, honeysuckle and ylang ylang.

Galbanum gives a leafy effect, and blends well with courmarin and jonquille.

Myrrh has a fragrance of wood, and is excellent in violet, white rose and lavender.

Oakmoss can be used with lavender, ylang ylang and courmarin.

Olibanum (frankincense) has a wood odor to be used only with very heavy fragrances.

Orris (from the iris root) can be used with carnation, cassie, rose, geranium or violet odors.

Storax is to be used most sparingly, but is an important element in narcissus, tuberose, and wisteria. It also blends well with mimosa, syringa, muguet, and some of the lily essences.

Vanillin (from the orchid) goes well with acacia, cassie, corylopia, fern, jonquille and lilac.

I have found the vegetable fixatives are successful in holding the fragrance, but they do not impart to the perfume the emotional excitement obtained when an animal fixative is used.

For that reason, I favor the animal fixatives. They give perfume a longer lasting fragrance and also give you more for your money!

Chapter 17

Methods Of Extraction

How are perfumes made? Only the perfume blender himself can answer this question, for the compounding of perfume has been a closely guarded secret since the very beginning of time.

The methods used in the blending of high grade perfumes are trade secrets or family possessions, zealously guarded exactly as the secrets of a nation should be guarded.

No amount of money could possibly buy some of the secret recipes and formulas. As an example, the formula of the famous German Cologne has been a closely guarded secret of one family for generations, and could not be purchased at any price. Remember, chemical analysis cannot reveal the processes, even if it can identify the ingredients. The treasured perfumes couldn't possibly be duplicated unless the secret is betrayed.

However, there is no secret connected with the many processes used and the methods followed in the extraction of the oils and resins which make up the perfumes of the world. Even so, much must be known pertaining to perfume before the ways of extraction and the many hours of work and concern involved may be thoroughly appreciated.

We know scents, as they occur in nature, may be divided into "true oils" and essential oils.

Ordinary oils are greasy, even lighter than water, and do not evaporate. They do leave a greasy mark or spot on paper, cloth and other substances.

Essential oils, however, do evaporate to a certain extent, but without leaving a greasy or oily mark.

Many flowers and plants owe their fragrance to essential oils, but many plants and fruits contain both the essential and ordinary oils. It is most important to realize that before a liquid perfume can be made these must be separated. A perfume which would leave a greasy stain on garments would be worthless to the perfumer. However, nothing is lost, for the true oils which are greasy are highly desirable for lotions, facial creams, pomades and other cosmetics.

There are several processes used to extract fragrance from flowers and other substances. They are:

1. Distillation, where a still is used.
2. Extraction, where a volatile solvent is used in a huge vat.
3. Enfleurage, where cold fats are used.
4. Maceration, where hot fats are used.

The essential oils may be obtained by great pressure being exerted upon a substance, breaking the tiny cells, far too numerous to count, which contain the fragrances. This is the method used in extracting the fragrances in lime oil, orange oil, oil of citron and many other oils used in the perfume industry.

Some oils are extracted from the fragrant substances by means of alcohol or other solvents. The iris flowers are treated in this manner. While many flowers, such as roses, rosemary, lavender, ylang ylang and scores of others are subjected to the distillation method.

We are told that by far the cheapest and quickest method of extracting fragrance is by distillation. We find it is used chiefly to extract oils from roses, lavender, orange blossoms and sweet scented geranium leaves. In fact, most flowers may be treated by this very old and simple method.

First, the petals must be carefully picked, while still wet with the dew in most instances. I have found 1:30 a.m. is the time! The new buds, with clean petals, should be selected. It is then necessary to weigh them.

In constructing a still it is necessary to find a clean tin oil can. The size should be one gallon.

Place the petals you have selected inside the can. Add eight ounces of distilled water (water not containing chemicals) to each pound of petals.

It is then necessary to find a cork which will fit tightly over or in the mouth of the can. Carefully make a small hole in the center of the cork. Through the hole in the cork run the end of a piece of copper tubing, which should be four or five feet long. There must be absolutely no space for air around the tubing in the cork. With putty or florist's putty seal tightly all around the tubing inserted in the cork.

The easiest (and safest) way is to build a fire in your

barbecue out in the yard. When the coals are sufficiently aglow, place a pan of boiling water on the grille; set the can of petals inside of this. Now the other end of the copper tubing must be placed in an open jar (not larger than one pint) and this jar must be as far away from the fire as possible, and at a much lower level than the can on the fire. Be sure the end of the copper tubing does not touch the bottom of the glass jar. Quite a large pan of cold water should be placed between the can of petals on the fire and the glass jar. The length of copper tubing must lie in this cold water.

It's said the true magic of this method of extraction is to be certain the copper tubing is coiled <u>three times</u> as it runs through the cold water. Another thing which must be checked very carefully is this: be certain the can containing the petals on the fire is the highest point, then the pan with the tubing coiled within the cold water should be on one level below. Then, on a level below the pan containing the cold water, should rest the glass jar holding the other end of the copper tubing. Doesn't it remind you of a vaporizer?

You see, as the water in the can containing the petals is heated more and more it vaporizes. It is most important that this vapor pass through the petals, for that extracts the fragrant oils, then it is chilled as it winds its way through the tubing, resting in the pan of cold water and condenses. Soon you will see tiny drops falling into the open jar. When you see those tiny drops you will know your experiment has been a success! Obtaining every single drop of precious perfume oil from the petals will take several hours.

Finally you will notice the tiny drops have stopped falling into the glass jar. It is then you will know your experiment is finished.

It is then that you remove the jar from the end of the copper tubing and place the jar in a cool place where it will be clean and free from dust or smoke from your fire.

After the contents of the jar have cooled for four or five hours, you will note the water has separated from the oil--allowing the oils to float on top of the water. Now comes the tedious part!

With a glass medicine dropper, carefully remove the oil from the top of the water in the jar, placing the oil in a dry sterilized bottle. It is better to select amber colored bottles which will reflect the light. The drops of oil in this tiny amber bottle are known in the perfume world as "Essential Oils." It is advised that they be kept in a cool dark place because they are "raw" and have no fixative to hold them constant.

In the commercial perfumeries these "cans" on the fire are large enough to hold hundreds of pounds of petals. Very elaborate equipment is employed; now, all this distillation is done automatically. The little still on the open fire described above is simply to have fun as you try your hand at making perfume.

There are many flowers so delicate and fragile they would not survive the harsher methods of extraction. Therefore, the fragrance from these most delicate flowers must be extracted by enfleurage or maceration, depending upon the ability of fats to absorb their odors.

Just as butter will absorb the odor of cheese or the other strong foods in the refrigerator, so will fats absorb the fragrance from flowers during the extraction process.

Enfleurage is the oldest method of extraction known to France, but is now only used when extracting the fragrance from the Jasmine and the tuberose. This is a most interesting process.

The petals of the flowers are spread upon glass sheets which have been coated on both sides with an odorless fat or grease which is roughened and slightly grooved. The sheets of glass are then placed between wooden frames and piled in tiers. In this way the petals are enclosed between the layers of grease.

The fragile jasmine petals must be replaced by fresh petals daily while the tuberose petals, which are more sturdy, are replaced only every two or three days.

The flowers are then removed by rapidly revolving brushes which are so set that they do not touch the grease, just very lightly lift off the petals. The tiny remnants which may stick fast to the grease are removed from the scent impregnated grease by means of a very special type of vacuum cleaner! In other words, this vacuum pulls the petals, or any remaining parts of the petals, from the grease without disturbing the fragrance in the least. The perfumed fat is called "pomade."

Practically all flowers, with the exception of the jasmine and tuberose, are treated by maceration. This method of extraction is exceptionally successful with roses.

When this method (maceration) is employed, the petals are placed in hot oils or liquid fats from 60 to 70 degrees centigrade. The flowers are stirred into the hot fats until the tiny fragrant cells are finally ruptured, then the fats are screened most carefully. This is then placed under pressure until all the fragrance is squeezed out of the fats and the same fats can be used over and over again, but only in the extraction of the same kind of flowers.

Only many years of experience can teach the perfumer the exact weight of the various flowers in proportion to the amount of grease which is necessary for the best results. There are no books to study for this training. One perfumer must teach another by constant observation and experience.

When the proper quantity has been treated the grease is given a final straining and the flowers are subjected to heavy pressure in order to squeeze out all perfume adhering or absorbed while in the grease.

Finally, the essential oils are separated from the fat or grease by means of solvents. Such solvents were unknown, we find, prior to 1856 and were not really successfully used unitl 1890. Prior to that time, some of the oils were extracted from fats by an alcohol wash. Even earlier than that the scent impregnated fats were used in their entirety just as they were removed from the press.

Perfumers have their problems, believe me. Quite frequently the extracted oils secured by the solvent method have an extremely dark color. However, when they are extracted by distillation, with paraffin wax, or by exposing the extracts to ultra-violet light, all this objectionable color is removed. Then everyone is happy!

The yield of fragrance obtained by these last two processes is very small in proportion to the amount of petals or blossoms used. Therefore, it is extremely expensive.

There are many other substances to be combined with the scent, some to enhance it, others to clarify it and still others to fix it, thus making the fragrance enduring and lasting. In doing all this, we know, every perfumer uses secret formulas, and tricks of the trade, which he has developed through years of practice; learned by trial and error only.

Until quite recent years, the method of extracting fragrance in India was extremely simple, and crude when compared to our modern methods, but definitely delightful!

As an example, they placed their rose petals in clay stills, weighing them as they did this. They they added twice their weight in pure water. The open vessels were left exposed to the air for an entire night. Believe it or not, they were able on the following morning to skim off the precious otto of roses.

This process of extraction took many rose petals, many days and nights of watchful effort so eventually there were those who just could not wait! Those eager beavers came upon what they thought was a clever discovery.

They added shavings of sandalwood to the petals of the rose to hasten the extraction. But, in their haste, they did not stop to realize the sandalwood was a much more potent, overpowering fragrance than the gentle perfume of the rose, so what happened? The entire batch, or experiment, was lost!

India has always been famous for its many oils. Quite often they added the gingelly oil seeds (also known as gingili) to fragrant flowers, placing them in alternate layers of crushed seeds, flowers or petals, then crushed seeds. (Always making sure the first layer and the last layer was the crushed seeds!) These were kept in covered vessels. As this mass of material aged, fresh flowers, more crushed seeds, even flowers bearing a heavier and more potent fragrance could be added from time to time to increase the intensity of the perfume. As this aged, the oils rose to the top and were skimmed off by the perfumer, then placed in tiny vials and sold for a handsome price!

The method I employ in making perfume without alcohol or any other base is quite different from those I have studied. I am a born experimenter; the methods of others have never interested me very much. As I read of how others did their extracting, I found myself knowing there must be a better way! (For me.)

It was then the thought came to me: surely within each tiny seed rests a plant, with a stem, leaves and a flower. Certainly that flower must have fragrance. It became my challenge to bring forth that fragrance, from the tiny seed, and make it live and bring happiness to the world!

While making a coffee cake one morning, I suddenly realized I had taken flour (the ground embryo of the wheat seed) to that I had added yeast (a digestive enzyme) then I had added a bit of warm water to set my "sponge" thus starting the dough.

Suddenly I realized, after covering the sponge to keep it warm, it would grow. If I did not knead it down it would soon overflow from the bowl, cover the table and fall to the floor. Why, in time, it could even fill a room! Where did all that growth come from?

As I prepared breakfast, I peeled a fresh peach, sliced it into a small dish as a breakfast fruit. I added one teaspoon of sugar. By the time I had made the coffee and made the toast there were two tablespoons of juice surrounding the sliced peaches. Now where did that juice come from?

It was the enzyme chain reaction in both instances that stimulated the growth, as one natural ingredient had reacted with another. How simple it all was! Why had no one ever applied this simple process to the making of perfume? I could hardly wait to begin my experiment.

The fragrance of an orchid, when first it opens, always thrills me. It was so different from the usual fragrance of a flower, so exciting, delicate and yet quite overpowering. I had always thought it was such a pity that this lovely fragrance only lasted a few moments. In fact, many declare the orchid has no fragrance! That is simply because they have never watched an orchid blossom open, making the world around it a much sweeter smelling place--for a few moments.

Realizing the vanilla bean is the seed of the wild white orchid which grows in the South Sea Islands, I ground a fresh vanilla bean with my little mortar and pestle until it was fine as face powder. This powder I placed in a sterilized bottle, and to that I added the powder from a digestive enzyme capsule which we often found most helpful to completely digest our food, and to prevent the discomfort of indigestion.

To this powder, I added seven drops of "juice" extracted from a night blooming flower. Why? Because the flowers which bloom in the night are controlled by the rays of the moon, making them catalystic in their reaction. They are wonderful activators.

The seven drops did not completely dampen the amount of powder I had in the bottle so I, instinctively, added fourty-two more drops, making forty-nine drops in all. (I have no idea why I did this during my first experiment, but now I know I was guided to do so.)

I then tightly closed the bottle and set it aside to see just what would happen.

And happen it did! Within two weeks there were several times the forty-nine drops in the bottom of the bottle! Unable to resist the temptation, I very carefully opened the bottle and sniffed. The fragrance was heavenly. It was far from a finished perfume, for the "yeasty" odor of the brew could easily be detected. However, I then knew I was definitely on the right road to a great discovery.

It was then I began gathering petals, leaves, stems and just about anything with a scent, even some bark from trees and started, in the same manner, many "brews."

From these simple little brews, I received an education second to none!

I found to dampen the first ingredients, I must use either seven drops or seven times seven drops because fourteen drops, twenty-one drops or twenty-eight drops just did not cause the brew to grow. It remained absolutely dormant. Why this exists

I do not know, I only know it to be true.

When a brew was slow to grow, and I wished to speed it up a bit, I found by holding the bottle containing the brew in my hot little hands for five minutes a day (which gave me a few moments for just thinking) the brew would be stimulated to much faster growth.

Others, which grew too fast, thus not having the potency I desired, I placed on the second shelf of the refrigerator for five minutes a day. This slowed down the growth and increased the strength and potency. Does this remind you of anything?

It is exactly what a housewife does to control the growth of a yeast product such as refrigerator rolls.

Many scientists, who have visited my lab, have exclaimed, "Why, that is magic!"

When I have displayed my brews to them, they couldn't believe what they saw. Still they do not think of making bread, coffee or cake rolls, even the juice which appears on sliced peaches, as having anything to do with magic. Why is this?

It is simply a matter of accepting as a part of life those things we have become accustomed to seeing during our lifetime. As a rule, everyone of us has watched our mother or grandmother make bread, slice fruit for breakfast, etc., accepting the sight as a very common part of life. But, it seems no one ever before thought of applying such a simple method to the making of perfume.

This "first" for me has caused international excitement, countless visitors and others who have come to my lab from faraway places to watch the wonder of it all, and a great joy of satisfaction now is mine.

When I was invited to be the guest of Dorothy Fuldheim, on color television, we made a "Television First." I made perfume from the seed of the wild white orchid (the vanilla bean) during the half-hour program. It was such fun!

Now let us get back to our brew. When the brew finally fills the bottle (which usually takes from seven months to two years all according to its ratio of activity) to keep the perfume brewed from overflowing, I transfer into another sterilized bottle all but one inch of the perfume and dregs of the brew. From that one inch or "starter" will grow another bottle of perfume. In time, I have discovered from one brew, or dregs, I can obtain as many as four bottles of perfume. No one can discern the first bottle from the fourth. They are exactly the same in every respect.

When the other methods of extraction are employed, this is not the case. Other perfumers realize their first run is the best run. Therefore, it is the most expensive. By the time the third run is "drawn off" the vitality and strength of the petals, leaves, etc., is spent. The third run is used for toilet water usually.

When I draw off the perfume from the dregs into a sterilized bottle, I must stop its growth or that bottle, too, will overflow. So it is, then, that I add a tiny bit of ambergris to the perfume which holds it in suspended animation, until it touches the warm flesh of the one who will wear it. It then immediately resumes its growth by uniting with the living enzymes of the body and lives all over again. Warmth and activity is all that is necessary to make the perfume "perk," or come to life. That is why it is always suggested perfume be worn on the pulses of the body.

After the perfume has been drawn off from the dregs, the stopper is firmly inserted, and the bottle of perfume is set aside to age. As this aging process takes place, the activity of the fixative (ambergris) upon the perfume tends to make it grow more mellow and definitely more exciting in attitude. After ninety days this perfume is ready to blend.

It must be remembered, perfumery is an art which requires years and years of experimentation, but also an uncanny knowledge of the properties and peculiarities of every scent and ingredient used. This knowledge is as necessary when making a single flower fragrance as when making a blend of many fragrances.

The greatest art of all, I believe, lies in the combining of various scents to make a perfect blend where one flower does not stand out and shout above the others or overpower another flower which has a more delicate scent. In accomplishing this, each perfumer employs his own little secret methods.

My personal approach to blending is far different from that of others. To me these gorgeous fragrances are very much alive. They are the life given essences uniting God and the flowers; God and the animals to whom He gave a fragrance. As I join the precious oils (and the perfume produced by the brew can well be called an oil for want of a better name) together in a blend, it is exactly like calling together a group of my dear friends for a charming evening of fun and relaxation. It is a joy for me to behold them link hands and become as one divine entity which, I know, will bring happiness to others.

As the many fragrances unite to become one, an entirely different and delightful entity often quite suddenly comes into existance. How many times I have been startled by the miracle before me.

Often I have been frightened by the feeling of standing quite apart, merely watching these wonderful and breathtaking changes take place before my eyes!

Watching each perfume blend take shape and come into being is a very sacred thing to me. It is exactly like watching the entrance of new soul approaching on gossimer wings of light.

I set about preparing a formula for a blend exactly as a housewife, filled with original ideas, would prepare a new and exciting recipe in her kitchen.

Just as she knows baking powder will cause her cake to

rise, and sugar will make it sweeter, so I realize the ability of each fragrance I have brewed, and the attitude each and every one will project. It is this attitude of emotional projection which is my goal.

When I begin to prepare a new blend, I select first of all a happy fragrance, usually made from a night blooming flower, for the rays of the moon tend to make a happy fragrance. In making my selection, I must be positive to select a happy fragrance which is catalystic in its purpose and attitude. From there I continue to build.

The formula may contain only three fragrances or, as my "Fantasy," it may contain 753 fragrances. The important thing is to be certain each single scent selected for the blend has a purpose. As I blend them together, I first blend a catalyst, then a synergist, then a catalyst and so on, but making sure always that the first ingredient is a catalyst and the last ingredient is also a catalyst. For that is the secret formula in the Bible. Remember? One catalyst to activate each synergist with an extra catalyst to activate the whole, making alcohol or any other base absolutely unnecessary.

When the many fragrances, or suspended brews, have been blended it is then that I add the animal fixative for excitement. This may be musk, castoreum or civet, depending entirely on just how wild, or exciting, I want the finished perfume to be.

To guarantee proper blending of the many fragrances, and

to defy the analysis of the curious, the greedy, I then employ the art of succession.

It seems the art of succession was discovered by Hippocrates, who is known to be the "father of medicine." I learned this art while studying homeopathy. Many believe this art to be the first atomic theory. It is supposed to fracture each and every atom, and every molecule, and in reuniting they never reunite in their original form or aspect.

It sounds complicated, doesn't it? Well, it is.

Furthermore, it is extremely difficult to learn. One must be tutored in the art. It cannot be learned from a book. Many hours of painful wrist, arm and shoulder muscles accompany its study. It must be done by hand!

The important thing is, as far as I am concerned, it has been well worth all the effort for never have any of my blends separated even after setting on the shelf for over thirty years. Nor has anyone--up until now--been able to analyze my blends enabling them to duplicate my perfume.

After applying the art of succession according to the ratio ascertained by adding the number of ingredients in the formula to the number of enzyme chain reactions involved in each, then selecting the common denominator, the blended perfume is set aside to age for thirty days.

At the end of thirty days, I open the bottle and compare it to the first bottle of that blend I ever made. Often, if the season has been extremely rainy and humid, the fragrances extracted from the petals or leaves will be less potent than the extract taken during the dry season. When this happens, at the end of the first thirty days, I quite often have to readjust the formula so that it compares perfectly with my control bottle of that particular blend.

After this comparison proves to be perfect, then it is necessary to succuss the perfume (according to the ratio worked out for that particular blend) and once again it is tightly sealed and set aside to age for thirty more days.

At the end of the second thirty day period, or at the expiration of sixty days, once again the bottle is opened and compared to the control bottle. Only on very rare occasions have I had to readjust the formula on the sixty day check. Then, once again, the blend is succussed, the bottle is tightly sealed and it is set aside to age for thirty more days, making a total of at least ninety days for proper aging.

It is necessary to age only a very few of the most difficult blends for more than the ninety day period after which the perfume is ready to bottle.

The perfume is so sensitive it can only touch glass or gold without changing its appearance and its fragrance. When filling the spill-proof bottles, tiny gold funnels are used and the perfume is dropped from glass medicine droppers. This can be a very long and tiresome task, for some fragrances are thicker and heavier. They go, "drip...drip...drip," taking ages, it seems, to fill. While others are lighter in weight and will go, "drip, drip, drip" so quickly the bottles fill up in no time at all!

The spill-proof bottles proved to be quite a problem to me at first! I remember inserting a tiny funnel in the very small opening of the bottle and beginning to drop, from the medecine dropper, the perfume into the funnel. Imagine my surprise when not a single drop of perfume went down into the bottle. It spilled out all around the edge of the opening. What a great disappointment that was for me!

During the sleepless nights that followed this discovery, I suddenly realized in that bottle was a vacuum. How could I break the seal? Then quite suddenly it came to me. I hurried down to my lab and blew into the gold funnels, warming them with my breath. I then inserted them into the spill-proof bottles which began to fill the bottles with perfume. They filled perfectly! One more hurdle had been cleared.

It is almost impossible for me to fill more than fifty bottles per hour, even with the fast drippers, so you can well understand why it is necessary for me to continue perfume making as a hobby rather than a business.

For some unknown reason, the ancients had all the time in the world. They did what they had to do slowly, carefully and often the hard way. But they did it well, a fact I have always admired.

Today time is so scarce that it has become the most precious thing in our lives! Business is based entirely on mass production. Quantity is the magic word, while quality rarely rates consideration. Well, that is not my cup of tea.

During my life, I have watched many businesses (especially fragrance businesses) flourish to the high heavens then, quite suddenly, fall completely. Why? Simply because the quality of their product did not stand up to their advertising or publicity department's propoganda.

Having observed this, I determined to try and experiment by making and <u>sharing</u> my perfumes in a manner quite the reverse of the accepted method of doing business today. It has been a world of fun!

I spend endless hours perfecting each and every brew and formula. I make the perfume entirely by hand, exactly as the ancients made it so many centuries ago. I then have gone one step further, because making perfume has been a labor of love for me, and I have never sold it. I have <u>shared</u> it at my cost with others through all these many years.

When the demand became so great that it was difficult for me to keep up the pace, I did not weaken the perfume in a mannen whatsoever, cheapen it in any way, nor did I cut any corners to save time and money. I simply explained my plight in an honest fashion. I found people to be very reasonable, and

and most understanding.

I explained I was only one person, I made the perfume entirely alone, and that I hand made it with extreme care. I made them understand I could only make so much perfume in a year, so the first person to contact me would be the first person to receive it. In comparison, money means very little when balanced out with the great joy I have experienced by sharing my perfumes in this manner.

To those who have enjoyed my perfumes for over thirty years comes the knowledge that the first and last bottle of that perfume are exactly the same. How many of your favorite perfumes have you found to hold up this well?

When alcohol is added to perfume it is for one purpose, and one purpose only; simply to make money for the perfumer.

Even though you may pay $100.00 for a bottle of your favorite perfume, just leave the cap, top or stopper off the bottle for twenty-four hours and see what you have left. You will have only a very few drops of a dark liquid in the bottom of the bottle. That is the perfume--the rest was alcohol. Pretty expensive alcohol, don't you think?

Many perfumers today use from 87% to 92% alcohol in their formulas, and they insist it is necessary to make perfume with alcohol to "preserve" its fragrance. Let me tell you this, alcohol does not preserve fragrance, it pickles it! The minute alcohol is added to any natural fragrance the enzyme chain reaction is immediately killed. It will never live again.

In the past, many offers have come to me, all leading to one ultimate state: changing my hobby to a business. As these offers have come to me, I have never closed a door. I always leave each door open just a tiny crack. Who knows what the future will bring?

The arguments, in favor of this change-over, have been many. Some say, "Hire help!" Sounds simple, doesn't it? Well, it isn't. First of all, where in the world could I find someone sufficiently versed in chemistry to recognize a synergist from a catalyst simply by smelling the petals, leaves or stems? This takes years of trial and error.

Where could I find anyone sufficiently trained to realize which of the four essential enzymes known to man should be used as a starter for the brews?

Each enzyme has its own little particular work to do. As an example, one of the enzymes digests fats. I have found quite often my sense of touch to be a great assistance for, as I touch a seed and find it to be oily or greasy, I immediately know I must employ the lipases to start a brew of that particular seed. It is the work of the lipases to "digest" fats.

When I wish to make a fragrance from a stem which contains elongated cells, then I know I must employ the cellulase as my starter enzyme.

The most difficult of all is to ascertain whether or not a petal, leaf or seed is a protein or a carbohydrate; they can be real foolers!

Then, too, one of the very first things I learned which was also important to the ancients was the fact that perfume could only be made in a <u>still room</u>. This was stressed constantly in all the few remaining records left in very old books.

I found this "still" had nothing whatever to do with the art of distillation, or corn squeezing! It simply means the air in the "blending room" must be absolutely still. This proved to be quite a problem.

Then, quite by accident, I discovered that growing plants in my lab was the solution. As they grow in the early morning sun filtered through an east window of my lab. The plants give off exactly enough oxygen for my needs and I, in turn, give off just the correct amount of carbon dioxide for them to continue on with their photosynthesis. It makes an ideal combination.

However, it took a little time to select proper plants to serve this purpose. It was a bit difficult to determine the correct number and type of plants, and a lot of encouragement was necessary to make those particular plants realize their one purpose in life was to grow in my lab! The array I now have is most interesting. Many types of plants have become quite happy and now consider my lab their permanent home. The only blooming flower among them is the Christmas, or Inch, cactus, and its flowers have no fragrance. I dare not have a conflict of interest, you know.

In preparation for the days I am to blend perfumes, from the finished brews, I close my lab to visitors. Finally, as I sit at my blending table, I can fairly feel the stillness of the air. As a final check, I hold a thread high above my head, between the growing plants and where I sit and wait.

If the thread moves to the left or the right, I know the time is not right for blending, and I must wait another day. If the thread does not move, however, then I know that is the day for starting work!

I have found the fragrances, especially when in the brewing stage, are extremely sensitive to outside odors. I do not smoke. When I am blending the brews, I must also maintain a very bland diet, no onions or garlic and such, because the delicate fragrances must be protected.

As an example, the breath of a smoker will curdle the delicate brew of the rare and precious white orchid brew. Yet, after my "White Orchid" perfume is perfected, the brewing cycle completed, and it has been succussed properly and aged sufficiently, a smoker can wear and enjoy the fragrancs as a perfume. But you would never in the world recognize it as the same perfume when the same "White Orchid" is worn by an ethereal type blonde who does not smoke. On the latter it would be heavenly!

All these simple things, plus the many more complex things, I have learned in over thirty years of constant study, research and experimentation for which I have never been paid a cent.

Why do I not have help? I have never found anyone qualified to take over my hobby, who would be willing to devote their time to making perfume solely for the purpose of making others happy. In order to hire a trained person to help me, their wages alone would make it necessary for me to double the cost of my perfume. Then only the very wealthy could afford the joy of having it for their very own. Surely this was not God's great plan. I feel certain He meant for each and every one of us to have just one tiny touch of luxury, unadulterated, with no commercial gimmicks.

"Why can't you train someone to do this?" they ask.

I searched for years for that "someone" but have found no one sufficiently interested. For many years, I have lectured on "The Romance of Perfume," "Perfume From the Bible," and "The Art of Making Perfume." This I have done in my spare time.

At one period of my life, I began to worry over the fact that my lecture contracts extend over three years into the future. What if something happened? The thought of disappointing those who were depending on my lectures greatly distressed me. It was then a very dear woman came to me and asked if she might help me with my lectures. She had been with the Play House in Cleveland for many years, but she was still young, and bubbled with enthusiasm. When she entered a room, it seemed the lights became brighter simply through the reflection of her smile!

I taught her "The Romance of Perfume" lecture. She adored

giving it! This gave me more time to work in my lab. But, evidently, this was not supposed to be for, quite suddenly, her life on this earth was over. How I have missed her through the years! However, we must face reality and know the world is a much better place just for having known the warmth of her smile. Working with her favorite perfumes, the ones she especially loved so dearly, has kept her with me always.

During all the years I have made perfume, there have been five I have thought would follow in my footsteps. Each time something happened to shatter that dream and erase that feeling of future security. Some felt they could not wait years for the brews to grow, others felt they wanted to make money faster than perfume. Still others decided they must smoke. It was then I realized I must continue on alone, and my search ended never to be resumed.

Today I am old. I do not have another thirty years to give, I do not have another twenty years to give...perhaps, not even one. Still I cannot bear the thought of leaving this life without there being left a few footprints on the sands of time telling where I have stepped in "The World of Fragrance." Therefore, with this desire in mind, I wish to give you this concentrated information describing my method of making perfume without alcohol or any other base.

When setting a brew, it must be remembered the ingredient which contains the fragrance is, in most instances, the synergist, the working element necessary to create the perfume. This may be the flower, the leaf, the seed or the root of a plant--where ever the fragrance of the plant is found to exist.

When this synergist is a flower it is most important to pick the flower at the proper time. The flower must always be picked between midnight and three o'clock in the morning, well before day-break. Should the flower reach its peak of maturity at the moment of the full moon after midnight is an extra bonus, because then the brew will grow more rapidly and be stronger, healthier and vital as it brews.

The flower must be cut on the diagonal of the stem as close to the bloom as possible. Immediately drop the cut flower into a sterilized bottle or jar (both quite cool.)

When leaves, which are fragrant, are used, they must be placed carefully in the bottom of a sterilized jar, but be careful not to pack them down one on top of another, for they should be free to breath.

When a root is to be used as the synergist, such as the iris root, this must be dug in the cool of the evening or the very early morning, washed and then very carefully scraped with a piece of glass to finely shred the root, thus letting the fragrance cells more easily escape into the brew.

After placing the above in the sterilized jars, sprinkle it with a tiny bit of the enzyme (a catalyst) indicated to activate that particular synergist. As an example: elongated cells require the enzyme cellulase, a carbohydrate, and most flowers, being carbohydrates, require the enzyme amylase. Proteins, in the form of seeds or beans, require the enzyme protease. A fatty synergist, oily to the touch, would interact with the enzyme lipase.

When using the finely ground vanilla bean (the seed of the wild white orchid) a combination of enzymes must be used: 1/3 cellulace, 1/3 protease, and 1/3 lipase to start the desired growth. The above enzymes must be obtained from marine vegetation which grows on the floor of the ocean--not seaweed or kelp which float in the water, thus becoming oxidized. This results in an impure brew or no brew at all. When searching for the enzyme powders, be certain they have been dehydrated by the "cold pressure method" removing only the water, leaving a dry powder obtained entirely without the application of heat in any form. This is most important!

To the above add either seven drops, forty-nine drops of seven times seven drops of juice squeezed by hand (not with any metal instrument) from a night blooming flower picked in the bud stage after midnight and before dawn. It is known that the night blooming flowers are controlled by the rays of the moon, rather than the sun, thus giving them a catalystic attitude. It is also possible at this point to use the same number of drops carefully extracted from a brew in progress made from the night blooming jasmine buds. However, the juice from the fresh night blooming jasmine bud is far better and makes a faster growing brew.

At this point, you have three ingredients in the jar: the synergist, the enzyme catalyst and the juice from the night blooming flower as an extra catalyst. Very gently swish all this around in the jar to let the drops circulate among the parts of the synergist. Then cap the jar air-tight.

At this point hold the jar in your two hands--concentrate on a happy growth within the jar--and hold the jar tightly and project the energy from your body throught the glass jar and into the starter of the brew. This should cause a bit of "steam" (even though your hands are relatively cold) to form within the bottle, thus denoting the first sign of growth of the brew. Let the brew rest and grow. The temperature should be as near to 70° as possible. Within a few days, should you not see tiny droplets of perfume forming within the jar containing the brew, hold the jar again in your hands for at least five minutes, again projecting your energy through the glass and into the brew, constantly urging the brew to grow beautifully, thus giving happiness to the world.

Strange as it may seem, these brews must be subjected to great love and devotion because they are extremely sensitive and require great encouragement to even make the first effort of growth! Urging them, in a very soft voice, is of great assistance to them. Even though all this may seem strange to you, remember it has really worked for me for over forty years! Some believe one must have a certain touch of magic, but I am inclined to believe one must have only the proper mental attitude. To be generous at heart, really desire to make others happy with your perfume, those are the real requirements. Greed, and the desire to commercial success and gain with aspirations for fame have never been my companions in "The World of Fragrance." Perhaps that is the reason it has been my escape from the world for over forty years.

Within six months, after starting the brew, sufficient perfume should be formed to completely cover the synergist used. Some brews grow far more rapidly than others. Do not set a time limit for them--be surprised!

I use eight ounce jars for my brews and, sometimes, within a year the jar will be two thirds full of perfume. Other times it has taken seven years to fill only a quarter of a jar! To open a jar, just to sniff the brew to see how it is doing, greatly slows down its growth. Sometimes the thrill of just one little sniff is worth the penalty.

By the time the jar is two thirds full, a portion of the brew may be extracted very carefully, by using a glass eyedropper, for experimentation.

When the jar becomes almost full, with danger of it running over, very carefully remove the perfume, with a glass eyedropper, so the dregs are not in the least disturbed. Be certain to leave the dregs in the bottom of the jar <u>plus</u> at least one inch of the perfume or whatever amount is sufficient to cover the synergist used to start the brew. This is most important. I have discovered during the last few years if these directions are followed to the letter the jar will fill with perfume seven times; but not, however, on the eighth time. After seven fillings you must start over, making an entirely new brew. Just amazing, isn't it?

The perfume extracted from the full jar should be held in suspended animation. The proper way to do this is to add a bit of ambergris from the sperm whale. This puts the perfume to sleep. As it quietly sleeps, it ages and becomes more mellow. Later on to awaken the perfume you succuss it by holding the jar containing the fluid firmly in both hands and proceed with the art of succussion. This art can be taught to you by any homeopathic physician, but it cannot be learned from a book. Remember the necessary strokes of succussion are indicated by the ratio of the enzyme chain reaction ratio employed in making the brew. After the aging process has been completed, to awaken the perfume, simply succuss it several times by sharply striking the bottom of the bottle or jar on the heel of the hand as you tightly hold the container in the other hand. When tiny bubbles form in the perfume you will know it is awake and ready to go to work for you.

This much must be remembered always: only use glass or gold to touch the perfume, never metal, even in the brewing stage. Stir, or separate the leaves, if necessary with a glass rod. Be certain no chemicals ever come in contact with the perfume, and be sure the air is free of chemicals or any strong odors. Never work with the perfumes after eating onions or garlic or fish, as they will pick up the odors immediately, thus ruining the perfume fragrance.

If your lab can be, at least, half submerged beneath ground level it is of great assistance. The ancients made their perfumes in caves, where the air was always still. Patience and perserverance are imperative attributes for the one who decides to make perfume "from scratch" the natural way, for it cannot be hurried under any circumstances. But when I say the reward, the finished perfume, is well worth it all, you can believe me.

It is my thought, hope and prayer that someone who reads this book will find the inspiration, as well as the help he or she would need, to unlock the door to the joys only to be found in "The World of Fragrance."

The verse in the Bible was the open door for me. I found the formula in the Bible at the very moment I needed it most.

From that moment on has come great happiness to me. Yes, extreme happiness for me and the thousands who have enjoyed my perfume with me.

For this I am thankful with all my heart.

Chapter 18

Infinite Guidance

Recently, I was asked, "Is there an astrological aspect to be considered when making perfume as the ancients made it?"

"Certainly!" was my instant reply.

However, quite often flowers do not reach their peak of perfection at the most propitious moment astrologically. This is most unfortunate.

Occasionally, my flowers have matured perfectly, the day has been overcast so their strength was not over-taxed, the cool moon lit night made it a perfect time to gather them, and the world seemed to be in perfect tune. When this task of gathering flowers was finished, I immediately set my brew (at 1:30 a.m.), as I have done many times before in previous years. Soon, though, I found the brew to grow far beyond my wildest anticipations!

Believing there must be a reason for this irregularity, I checked and found that at the time I gathered the blossoms there was a full moon. Could that have made the great difference? Who knows?

The night blooming flowers, which are controlled by the rays of the moon, are the strongest catalysts. The aspect was there.

The very next year, the same plants allowed their flowers to reach complete maturity. At exactly the proper time and lighting I gathered the blossoms. Once again the brew was set, just as before. But, the moon was new!

The depth, the charm, the emotional value of the resulting fragrance, after brewing, all existed exactly as in the previous year--but, the brew grew only half as fast. The great strength of purpose, the strong vitality seemed to be lacking. Just another question mark, another challenge in the World of Fragrance.

How I am inspired to design, to originate, a new perfume! One may ask, "Do I really make perfume from scratch?

The answer is "Yes!"

Making perfume from scratch, from seeds, petals, leaves, stems, bark and roots is the real challenge in the perfume world. I cannot possibly imagine experiencing any satisfaction whatever in simply attempting to copy the work of others.

How am I inspired? In answering this question, I hope I may convey to you the tremendous impact from which is derived my inspiration. It can only be compared to the driving rain against a windowpane! It is both startling and persistant, and never to be ignored.

The inspiration may awaken me in the middle of the night. I have often awakened believing so strongly I was surrounded by flowers that it has been necessary for me to turn on the light beside the bed to be certain there were no flowers on the bedside table. Of course, there were none! Nevertheless, <u>for me</u> the air was filled with a heavenly fragrance of many flowers in perfect combination.

The combination of fragrances has always remained with me for a sufficient length of time that I have been able to capture it in a new and lovely perfume. It may take me years to discover the exact combination, perhaps as long as seven years, but eventually the discovery is mine. I am able to recognize it instantly, the moment I have found it! I rejoice with a thankful heart.

At other times, the inspiration for a new fragrance may come to me as walk quietly in the woods. Perhaps, as I work in my garden, the feeling will be overpowering--the fragrance which comes to me is so compelling. No matter where I may be, asleep or awake, the feeling is always unmistakable. I have come to recognize it at once.

Several times such inspirations have come to me when I was thousands of miles from my lab. The first time this happened to me, I was filled with consernation. For just one second I was overcome with panic. It was then I began to sketch my thoughts with pad and pencil. I found to my amazement the fragrances were, and always had been, accompanied by the vision of a girl.

This girl is dressed in keeping with the fragrance, be it Egyptian, Persian, or modern. Always fair of face and form, she is the effervescence of fragrance personified! Not only young, but always vital, she is poised and self-confident, and as beautiful as a bird about to take off in flight to parts unknown!

As I now look back, I can once again see each and every one of them as clearly as living things, and always in vivid color.

One evening at twilight, far from home, such a vision appeared to me. A lovely girl walking in the garden. She stopped to gather flowers to place in the basket she carried on her arm.

Pad and pencil were in my hands, for I had been making notes of an experiment I had planned to make when I returned to my lab.

Rapidly I made notes, sketching each bud, each flower, and every leaf as she gathered them. Finally, her basket filled, she just seemed to fade away into the twilight, leaving behind her a fragrance which could not be forgotten. Nor could it be denied.

Each day that followed, I studied the sketch thoughtfully, at which time the new and distinctive fragrance always returned to me exactly as in the beginning.

As soon as I returned to my lab, simply by referring to the sketch, I found I was successful in capturing the fragrance in its entirety.

I named this first perfume, which came to me in this manner, "Dream of the Night." This name is derived from the fact that it is derived from flowers which only bloom after dark. The sketch plainly portrays the girl picking the flowers, the buds and the leaves by the light of a rising full moon!

At other times, a story I have found in ancient history has

served to furnish me with the needed inspiration. An ancient custom of the orient, known as "The Ceremonial of the Jade" filled me with the desire to try to duplicate, in a perfume, the feeling of great exaltation experienced by those who attended that very special event.

History told me the ceremonial was held surrounding a god made of solid jade. All the flowers used at this particular ceremonial were completely white!

The orientals believe the jasmine bud symbolizes the purity of womanhood, thus the at the Ceremonial of the Jade the jasmine bud predominated. Few people realize the delicately beautiful, faintly fragrant blossom given to us by the jade plant.

All this came to me in a flash as I closed the book upon this most interesting story.

I began working at once. The result was the combination of jasmine bud, the precious jade flowers, plus the other white flowers used during the Ceremonial of the Jade by the ancient orientals. I called this discovery my "Jasmine and Jade." It is very, very different. Its depth is purely oriental, and its memories are most lingering. Better still, it can be worn beautifully by heavy smokers.

For my first experiment of "Jasmine and Jade," I had a tiny flacon containing only a few drops of the pure essence of the jade which had been sent to me by a dear friend who is an importer. It had to come from China.

Suddenly I realized a supply of this ingredient would no longer be obtainable! I set my experiment aside to age knowing, if I could only be patient, the answer would come.

Months later, while visiting a dear friend in Pasadena, California, we were walking in her garden. A clear, interesting fragrance filled the air. I looked up to the top of a thick hedge, and on the very top were the fairy-like blossoms, the perfect stars of lace, which could only burst forth from the jade plant!

The next day, as the jet brought me back home, my very large purse was filled with slips from this hedge as well as its delightful blossoms. The slips are all growing beautifully, and the blossoms were more than enough to start two brews from which I have had sufficient perfume essence to make my "Jasmine and Jade" for all who have desired it for their own.

Of one thing I am certain: it has been absolutely necessary that I have complete faith in the guidance I have obtained and constantly known to be with me, or I could never have known even the slightest degree of success!

When one realizes the addition of one tiny ingredient, not intended to be present in a formula, could quite easily destroy the work of years in the World of Fragrance. This Infinite Guidance simply has to be recognized!

How do I know these things to be true?

How do I know that I am a woman?

How do I know that it is day and not night?

How do I know tomorrow another day will dawn?

It is by instinct, by Infinite Guidance, that I know these things to be true.

It was through desperation I learned the lesson of Faith.

Only through Faith was I taught to accept Infinite Guidance as my personal need.

Chapter 19

Incense

Of all the perfumes known to man, the most simple consists of powdered flowers, fragrant leaves, woods and the aromatic resins which are designed and prepared to be burned, giving off their lovely fragrance in the form of smoke.

This type of perfume is known the world over as incense.

It has constantly been used in all types of religious ceremonies since the very beginning of time.

Some even believe the discovery of such incense probably came about when the fragrance or aromatic odor of certain woods used in the camp fires pleased and attracted the attention of primitive men. It is easy to believe the particular wood or leaves giving off the odor would have been quickly recognized or identified by the primitive man who would continue to burn them because of their pleasant aroma.

As these men of the earliest ages always shared their best and most pleasant things with their gods, they soon came to burn the selected woods and leaves as offerings to gain the favor of the gods they worshipped.

This thought and desire had become so deeply rooted within man that this custom has carried down through all the ages, for incense is used even to this very day in our churches.

It is also known that even today some savage tribes burn incense, which they prepare from aromatic woods, herbs, flowers in ceremonies designed for the purpose of scaring away their devils! It is amusing to learn their reasoning in this. They believe any scent which would please their gods would, quite naturally, displease their devils. Of course, we have no idea how successful they were in ridding themselves of their devils, but we do know that even today we are using incense for fumigation purposes and to rid ourselves of mosquitos and other insect pests. To me, they are quite one and the same!

Without a doubt, the Chinese use more incense than any single nation.

They use incense for many purposes. Their temples fairly reek with is almost overpowering fragrance.

Every devout Chinaman is supposed to burn three sticks of incense each night and morning to his gods. If he desires special favors, he then increases the number of "joss-sticks."

All ceremonies, weddings, and funerals are accompanied by "joss-sticks" beyond number!

One of the fondest memories of my childhood is renewed each time I light a stick of punk (incense) given to me by one of my friends who knows how I adore it. This delight can be purchased only in oriental stores, as a rule, for they alone appreciate the great feeling of relaxation which accompanies the burning of incense.

It is well to remember that any material which will produce a fragrance when burned may be considered an incense. The use of incense is one of the very oldest methods of producing household fragrances. When you place twigs or stems of rosemary, acacia, aloe, lavender, mignonette, thyme, southernwood and sweet fern on the burning coals of your fireplace, you are producing a famous perfume, in the form of incense, which was used thousands of years ago.

One of the most delightful adventures of a housewife is the making of her own incense. Many herb's which grow in our garden can be dried, powdered, and burned in our homes, giving the entire house a most interesting and "homey" scent. This adventure holds many surprises. When orris is dried and burned it gives off a most pleasant fragrance of violet. One would declare there were freshly picked violet blossoms in the room!

The Chinese are extremely fond of the fragrance given off from burning southernwood; when the leaves are dried and powdered and then burned, tobacco odors are immediately destroyed.

Just a very tiny pinch of powdered sage, or southernwood, on the stove will completely dispell any cooking odors. It is well to keep some of these herbs dried and powdered in a small air-tight jar near at hand for instant use.

Incense may consist of merely a few herbs, or it can be a most complicated and exotic blend just as a perfume.

There are many ways in which dried flowers, from your flower beds, or herbs, from your garden, may be used in the home. Incense, potpourris, sweet bags, linen pads, scented cushions, sachets, bath preparations, are all pleasant hobbies.

Any of these may be made quite easily and most economically right in the home. When carefully prepared, and nicely put up in attractive packages, they are lovely gifts which will be welcomed everywhere. These are also excellent little money-making ideas for clubs, and make wonderful group activities. Why don't you try it sometime?

Today, I believe, incense is taken far too lightly. The ancients recognized its supreme emotional value and capitalized on it to their advantage. Books of great length have been written about incense. Yet, how many of us in this modern age ever think of employing it in our homes or our business. Much could be gained by this simple addition of incense to our every day lives.

You see, the use of incense is perfectly scientific. All occult students are aware of this one fact: "There is no such thing as really dead matter." Every single thing in nature possesses and constantly radiates its own variation and vibrations. Therefore, every chemical element has its own set of influences which are useful in certain directions and, we might say, harmful in others.

So, thinking in this way, it is quite possible to mingle certain gums or resins which, when burnt as an incense, may strongly stimulate the purer and higher emotions. However, a mixture could be made just as easily which would promote the most miserable of feelings.

The reactions caused by inhaling the fragrant fumes of incense, so well known to the ancients, would be a life long study for any one of us today. But this we do know.

Incense has existed down through all the ages. It has been constantly used throughout the centuries in the churches and temples, and it has one decided advantage: it rises into the air, and wherever a single particle goes, the purification and blessing is borne with it.

We find practically all the religions of the world use incense in one form or another. We find it in the temples of the Hindus, the Zoroastrians, the Jains, and in the Shinto temples of China and Japan. It was used in Greece, Rome, Persia and in the ancient ceremonies of Mithras. All these people, including the Roman Catholics even today, avail themselves of it. Why? Because they realize it is a most valuable and useful thing.

At one time in history, incense was known as "Perfume used for Fumigation." We are told, according to the use made of them, perfumes for fumigation may be divided into two groups: those which develop their fragrance upon being burned, and those which do so on being merely heated. It is the most interesting to know the former group includes pastils and ribbons while the latter, fumigating powders and waters. Let us consider each.

Pastils consist for the main part of charcoal to which

sufficient salt-petre is added to make the lighted mass glow continuously. This leaves pure white ash. To the charcoal and salt-petre are added various aromatic substances which are gradually volatilized by the heat. As the mass continues to burn, the air is filled with perfume, giving happiness to all who inhale the fragrance.

It is of interest to note that only ordinary salt-petre (nitrate of potassium) can be used for this purpose. The socalled Chili Salt-petre (nitrate of sodium) which becomes moist in the air cannot be used. For many pastils very finely powdered fragrant woods such as cedar are frequently employed. As this starts to slowly burn, the wood gives off products of a pungent and, sometimes, disagreeable odor. These malodorous substances includes acetic acid which greatly lessens the fragrance. This must be taken into consideration. The really fine pastils are composed of resins and essential oils which are usually formed into cones less than an inch high either in the hands or by pressing the substance in metal molds.

Each solid ingredient, when making fumigating pastils, must be finely powdered by itself, and then the necessary quantities are placed in a wide porcelain dish and mixed with a flat spatula of wood. Never metal.

During this operation, to confine all the precious dust, it is wise to cover the dish with a cloth, working with the hands under the cloth. When the mixing is completed, the essential oils are then added. Alternately added, at that time, should be sufficient mucilage of acacia to form a putty like mass which must be kneaded with a pestle. After drying this will form a very firm consistence.

There is a very old, and most famous, formula known to all perfumers for centuries, often printed, which reads "Pastilles Orientales. It requires 1 and 1/2 pound of charcoal, three and 1/2 ounces of salt-petre, one pound of benzoin, three and 1/2 ounces of powdered amber, and two and 1/4 ounces of tolu balsam."

The charcoal used for this formula, and all other pastills, should be made from soft woods such as poplar or willow.

The outstanding characteristic of such pastils is the amber they contain on ignition gives off a very peculiar odor much prized in the orient, but not so favored in Europe or in America. Therefore, when making pastils for sale in Europe or America the powdered amber was replaced by a powder made of crushed petals, dried crush leaves, etc., which gave off a more delicate fragrance as it burned.

A mortar and pestle (or wooden mixing bowl with a large wooden spoon or wooden potato masher) is used to make a powder from the herbs, leaves, stems or petals. The best results are obtained when the material to be used have been thoroughly dried first.

I have found this drying process can be quite simple if

a wire cooling rack (such as those used to cool freshly baked pie, cake or cookies) is covered with one layer of paper towel and the herbs, etc., are spread out in a very thin layer over the paper, allowing them to dry at room temperature. In this way, air is allowed to circulate beneath the paper covering of the rack, thus the drying is accomplished without heat being applied. It is my belief that drying the herbs, leaves, etc., with additional heat, such as the "slow oven method" employed by some, destroys a certain amount of the fragrance. This is learned when drying herbs such as parsley, sweet basil and mint for food seasoning. It is well to remember the drying materials should be carefully turned night and morning as they dry for three days.

The steps in making incense are quite simple, but must be followed exactly for perfect results.

All dry ingredients must be reduced to a fine powder, then mixed and sifted (to be certain there are no pieces of stem or large particles present) then slowly stir in the oils. As this rests a few moments, dissolve the potassium nitrate and the gum tragacanth in water and add to the mixture. Rub to a stiff paste and, as quickly as form into small cones, tiny cakes or blocks, then allowing it a sufficient time for drying.

It is quite impossible to say exactly how much gum tragacanth should be added to the mixture. It is to be used in sufficient quantity to make a stiff paste which can be quite easily molded like putty in the hands. When only 1/5 of an ounce of tragacanth gum is dissolved in 1 and 1/2 ounces of water the resulting mass will be sufficient to mold at least three ounces of the mixed powder into cones, cakes or blocks.

The formula for making incense is quite simple and can be easily varied to meet the desires of the perfumer. There are a few such formulas which are known to many, and some can even be found in the library. Some, I have found to be most satisfactory and are as follows, "Sandalwood Incense--1/2 ounce of benzoin gum powder, 1 ounce of sandalwood powder, 1/2 ounce of cascarilla powder, 1/8 ounce of vetivert powder, 1/8 ounce of balsam of tolu, 1/8 ounce of potassium nitrate, and 1/8 ounce or as needed of tragacanth gum."

In a very old French book, I found the following formula for making incense. It is different, yet similar, to our more modern formulas. "Poudre d'Encense (Incense Powder)--1/2 pound benzoin, 1/2 pound cascarilla, 15 grains musk, 1 pound sandalwood, three and 1/2 ounces salt-petre, five and 1/2 ounces vetivert root, 1 pound olibanum, and five and 1/2 ounces cinnamon. Dissolve the saltpetre in water, saturate the powders with the solution, dry the mass, and again reduce to a powder."

This French incense is much different than other formulas I have experimented with. This powder was, in olden times, strewn on a warm surface (such as the top of a stove) at which time it would take fire spontaneously and gradually disappear, leaving a lovely fragrance in its wake.

The early French also had another most fascinating custom. They made "Fumigating Papers and Wicks" (Bruges Ribbons) which were known in France as "Papier a Fumigations " or "Buban de Bruges." In Germany much the same type of perfuming was carried on by their "Raucherpapiere" or "Raucherbander."

A very old book tells us the fumigating papers were strips of paper impregnated with substances which become fragrant upon being heated, but such a strip need merely be placed on a stove (or hot stone) or held over a flame in order to perfume an entire room. It seems these fumigating papers were divided into two groups: those meant to be burned entirely, and those meant to be used repeatedly. The former were treated with aromatics and then dipped into salt-petre solution. The latter, in order to be assured they would be incombustible, were first dipped into a hot alum solution so that they were only charred by a strong heat and never entirely consumed.

From this very old book--long ago out of print--I take the liberty of quoting the following formulas simply because I believe them to be of tremendous interest to anyone who, thus far, has enjoyed a visit in the World of Fragrance.

"Inflammable Fumigating Paper (Papier Fumigatoire Inflammable--the paper is first dipped into a solution of 3 1/2 to 5 1/2 ounces of saltpetre in water; after drying it is immersed in a strong tincture of benzoin or olibanum and once again dried.

An excellent paper is made according to the following formula: 5 1/2 ounces benzoin, 3 1/2 ounces sandalwood, 3 1/2 ounces olibanum, 150 grains oil of lemon grass, 1 3/4 ounces essence of vetivert, and 1 quart of alcohol. For use, the paper is touched with a red-hot substance, not a flame. It begins to glow at once without bursting into flames, giving off numerous sparks and a pleasant odor."

The "quart of alcohol" would be a king size problem today when attempting to make this formula. Grain alcohol can be purchased only on prescription, you know, and the price is sky high. It is not possible to use our "rubbing alcohol" for that contains chemicals to discourage humans from drinking it. Those chemicals could completely ruin either the perfume or formula such as the one quoted above. In the old days, pure alcohol must have been quite plentiful, for you will note it was used--in lesser quantity--even in the non-inflammable papers. I quote such a formula: "Non-Inflammable Fumigating Paper (Papier Fumigatoire Permanent)--this paper is prepared by dipping it in a hot solution of 3 1/2 ounces of alum in one quart of water; after drying, it is saturated with the following mixture. Seven ounces of benzoin, tolu balsam, tincture of tonka, essence of vetivert and 20 fluid ounces of alcohol. This paper, when heated, diffuses a very pleasant odor and can be used repeatedly. It does not burn, and only strong heat chars it. Some manufacturers make inferior fumigating papers by dipping the alum paper simply in melted benzoin or olibanum."

The French, many years ago, learned to make fragrant powders which were placed on hot stones (soap stones seem to have been useful for this process). As they burned they gave off a delightful fragrance. We are told the stones were carried by servants from room to room, thus perfuming the entire home. After considerable searching, I finally was able to find a formula for this most delightful powder. Today it is extremely expensive to make, but the fragrance obtained is sufficient reward for the time effort, and great expense. This is the formula: "Poudre Imperiale--3 1/2 ounces benzoin, 1 3/4 ounces of cascarilla, lavender, rose leaves, sandalwood, and cinnamon, 3 1/2 ounces of olibanum and orris root, 75 grains of oil of lemon, 30 grains of oil of clove, and 15 grains of patchouly."

It is well to remember, the imagination can be completely unharnessed when making incense! It may consist of only a few herbs, or it can be as exotic as a heady perfume.

The many stories which have been passed down through the ages concerning incense and its effects upon humans are most interesting.

Only recently I was amazed to find this little green gem in the comic section of our Sunday newspaper, "Incense for all the world's churches that use it is made by a Dutch firm from a series of ancient mystical formulas set out in the Bible, using ingredients from the same places King Solomon used to obtain spices. It was originally burned during pagan rites to overcome the odors of the sacrifices."

The colored illustration accompanying this bit of information is simply terrific! It shows a man (always a man, remember?) in ancient garb grinding with a very large mortar and pestle, the various sticks of spices and ingredients. Below is another picture of an altar whereon rests a lovely container which allows smoke to slowly curl out from the burning incense!

This has made a delightful addition to my collection.

Chapter 20

Fragrances Of Famous People

Alexander the Great--We are told, when the youthful Alexander was being tutored by Leonidas, he appeared wasteful by burning too much incense in his sacrifices.

The tutor reproached him by saying it would be time for him to be so extravagant in his tastes after he had conquered the countries producing the lovely frankincense which was the favorite scent of Alexander.

This immediately curbed the trait so emphatically that for many years Alexander affected to despise all fragrances.

In fact, when Darius was vanquished, Alexander ridiculed a gem encrusted casket filled with precious perfumes found among Darius' most prized possessions.

However, when Alexander finally conquered Arabia, he recalled the words of Leonidas and sent his old tutor an entire cargo of frankincense and myrrh as a gift.

From then on, Alexander surrounded himself with perfumes! We are told he constantly had his floors sprinkled with perfume while aromatic resins and myrrh were always kept burning in his presence.

Anne of Austria--Anne of Austria was known as the Queen with the beautiful hands. This fact is mentioned countless times in the history of cosmetics.

She was especially fond of perfumed gloves, and owned many

pairs.

She believed only the perfumers of Spain could properly prepare the leather, and during her reign perfumed gloves made of mouse skin, and imported from Spain, were most fashionable.

History tells us Anne of Austria was one of the first ladies to perfume her hair, and her fingertips!

King Charles VI of France--Have you heard of "Oiselettes de Chypre?"

These were artificial birds made of silk and feathers stuffed with perfumed powders which King Charles VI had made for his beautiful Queen Isabeau of Bavaria.

This started a fad which rocked the perfume world to its very foundations!

Circe--Circe, who ingeniously wrought charms to detain Ulysses on her island, used perfume as her secret weapon.

The story goes that her perfume was so potent it completely dulled his sense of duty, so he remained at her side.

Perfumes at that particular time were made in three types: Solids (unguents) which were from almond, rose or quince; liquids, which were compounded from flowers, spices, gums (resins) and oils; and powdered perfumes, such as our after bath powders, perfumed talc and the like.

All were heavily laden with musk and the more delicate animal fragrance of ambergris. Both are quite overpowering and persuasive under the proper circumstances!

Cleopatra--The great love that Cleopatra possessed for perfume was so overpowering that it deserves mention in this chapter.

Why she wore perfume, where she wore her perfume, as well as the tremendous results she obtained with this method will be found in the previous chapter concerning perfume in Egypt.

To me, Cleopatra, perfume, and Egypt have always been synonomous. So much so it was most difficult to decide just where to place the stories of her fragrant life.

Queen Elizabeth I of England--The very first royal gift recorded in English history was a pair of scented gloves, made from mouse skin, which the Earl of Oxford brought from Italy for Queen Elizabeth I.

At the time he made the purchase, the Earl also selected several other novelties in the way of scented clothing; sweet bags, a scented leather jerkin for himself, as well as several new scents were among his purchases.

Queen Elizabeth I became so extremely interested in perfume, after receiving this gift, that she sent the Earl back to Italy and Spain. She requested him not to return until he could bring her all the information and instructions pertaining to the making of perfume.

This was quite a task for the Earl, for the secrets of perfume have been well guarded from the beginning of time. He remained in his search for so long that the Queen became impatient and finally sent for him.

When he returned to England, he was positively loaded with precious perfume of all types, and he also brought her all sorts of perfumed clothing. But he was forced to inform her that he could not beg, borrow or steal the actual knowledge necessary to make perfume. He also explained all the legends he had learned about the Priests and learned ones who devoted their lives to the art of making perfume, which impressed her greatly.

To save his own neck, he did his best to convince Elizabeth I that those who devoted their lives to making perfume were Godly, dedicated people. He went into great detail explaining that much magic was connected with the art of fragrances. All perfumers spent much time in meditation, and he said many even believed their instructions came from God.

Queen Elizabeth I drank in every word, and gave the matter great thought. Soon she had built in her castle a still room which was a laboratory of sorts. There, surrounded by her various oils, she could retire to hours of deep thought and meditation. Believing she would, in this manner, "receive" instructions for making perfume.

Evidently, she soon grew tired of waiting for the secret to come to her, for we are told she soon began to experiment on her own. She came forth with several different, and quite unusual, perfumes as a result. Only one recipe has been handed down to our day. We are told this was her favorite and, upon her insistance, it was worn by her entire court--pour souls!

This was the formula, and you can judge for yourself: "Thoroughly crush very sour apples, add the fat from a young dog, properly blend and allow to age until ready."

Can you imagine rotten apples plus rancid fat...well aged? It must have been horrible. Down through the ages many perfumers have tried to make something, in the way of perfume, from this formula, but it always seems to come out the same; very dissatisfactory, to say the least.

When confronted with these results, the English reply quite calmly, "Most unfortunate! How sad that the Queen neglected to state whether the fat was to come from a male or female dog! Too bad, too bad!"

Nevertheless, Queen Elizabeth I did much to further perfume in her day.

Her skin was described as "white albion rock." To preserve this paleness, she used a lotion of her own making. It contained white of egg, powdered eggshell, alum, borax and white poppy seed. This was mixed with water and beaten until the froth stood three fingers deep.

She also washed her hair in pure lye. A compound of lye is made of wood-ash and water.

"Queen Elizabeth's Perfume" was made especially for her of majoram. She preferred a very light scent, nothing too potent.

We are told whenever books were bound especially for her,

the binders were warned well in advance that they should not use the customary leather of that day, which was treated with oil of lavender, for the scent was far too strong and heavy for the Queen. They were commanded to use only majoram.

She so adored flowers that she was surrounded with them constantly. During the winter months when fresh blooms could not be obtained, she had artificial flowers made for her and perfumed them with rose, honeysuckle, cinnamon and ginger. Even snowballs presented to the Queen were scented with rose water for her pleasure.

As she neared middle age, she became more and more interested in perfumes and cosmetics. She was her own best advertisement, for she was known far and wide for her youthfulness.

In 1599 great gifts of jewelry and robes were sent to her by the mother of the Turkish Sultan in exchange for "distilled water for the face and scented oil for the hands" such as she herself used daily.

This was not all, however: Queen Elizabeth I was the first one to keep the royal linens in scented chests!

Those chests were imported. Many were made of sandalwood or red cedar. She delighted in making little "scent bags" which she wore among her clothing and hung with her gowns.

She and her guests were constantly delighted with the glorious fragrances which wafted forth from everything touched by Her Highness.

Queen Elizabeth of Hungary--In 1370, Queen Elizabeth of Hungary first distilled what she called "Hungary Water." How happy she would be to know that it is still made and enjoyed by many even today!

The recipe was said to have been given to her by an old hermit. She guarded it most carefully throughout her entire life, because she believed it did wonders to preserve her great beauty for which she was famous.

There is certainly no argument on this score, and her beauty must have been quite overpowering.

At any rate, she was asked to marry the King of Poland when she was seventy-two years of age!

Her famous elixir is said to contain, "Rosemary, verbena, peppermint oil, triple rose water and triple orange flower with the addition of 90% of alcohol (which we are inclined to believe was ancient aromatic vinegar rather than alcohol as we know it today). Allow the entire recipe to age for six months."

Helen of Troy--Helen of Troy's fatal beauty, potent enough to launch a thousand ships, was attributed to the use of perfume compounded especially for her by her magicians.

In other versions, we are told she received her beauty secrets from Venus, the highest authority of all!

Nevertheless, she spent many hours each day relaxing in her scented bath. After which, she commanded her handmaidens to anoint her entire body with many fragrances.

One special perfume scented her hair, another her ears, another her throat, and so on until the ritual ended with perfuming her lovely toes!

What a luxury that must have been!

Empress Josephine--Napoleon's Empress Josephine was of Creole blood, so she preferred the heavier and more potent perfumes of her day.

She went so far as to have all of her favorite perfumes brought to her from Martinique.

It is also written that her boudoir at Malmaison was so completely saturated with musk that after sixty years of concentrated effort, trying to rid the place of the scent, the new owner found it positively unbearable--almost suffocating.

The new owner wrote quite an account of it, saying the very walls seemed to be impregnated with musk!

Many years later, it was learned, to attain this end result, Josephine had commanded her servants to completely cover the walls, the ceiling and the floors with many coverins (or coats) of pure musk.

The consumption of perfume was never greater than it was during her reign. The perfumers adored her because she allowed them to live on the fat of the land.

What a wonderful customer she must have been!

Napoleon--Napoleon's favorite fragrance was the violet.

Because of his great passion for the violet and its perfume, the House of Bonaparte chose that flower as the emblem for their house and crest.

Throughout his life, he enjoyed both the flower and its fragrance so much he kept himself surrounded with one or the other constantly.

It is said the court perfumer supplied him with two quarts of violet perfume each week. This he used to bathe his head!

Those who loved him wore violets or hung bunches of violets in their homes in his honor. Nothing could have pleased him more.

We are also told as Napoleon lie dying in exile, lonely as he was, he constantly fretted to the very end over the delay in shipment from France a dozen pairs of violet scented gloves he had placed on order.

They arrived too late.

Sir Walter Raleigh--During the reign of Queen Elizabeth I of England, Sir Walter Raleigh was the dandy of her court.

In fact, it was he who set most of the fads of that day. For him, nothing was too elaborate, too expensive, or too sensational.

Upon his return from a voyage to Spain and Italy, he wore a perfumed leather jacket, or jerkin, which was the ultimate of luxury in those countries during that period.

The Queen was so taken with this idea, she sent him back to purchase for her a full length scented leather coat which was one of her most prized possessions for the rest of her life.

When Sir Walter Raleigh sailed for the New World, he wore his perfumed leather jacket. We are also told he took with him "scents" with which to re-perfume his leather jacket. These were, evidently, the first perfumes to be introduced to the New World. I have been unable to find any earlier mention of perfume there.

When I read the account of this voyage, I felt it was well worth a bit of concentrated research. I was repaid for my efforts by learning this.

It seems Sir Walter was a most athletic man, full of pep, strength and vitality. He also perspired profusely!

The irony was that, as he perspired, the leather of his precious jacket became saturated and as a result, the odor of his perspiration far surpassed the more delicate fragrance of the perfume with which the leather had been cured!

Consequently, Sir Walter would have to frequently stop and re-perfume his jacket to overcome this disadvantage.

It was most inconvenient when in battle, so finally he discarded the whole idea.

However, he had made a name for himself in history as the first man to wear a perfumed leather jacket in the court, into battle, and on voyages of British State.

Madame Tallien--Fashionable ladies under the Directoire revived the perfume baths of the ancient Greeks and Romans.

Madame Tallien went even further!

She took daily baths of strawberries and rasberries, then had her body servant gently rub her entire body with soft sponges soaked in a mixture of milk and rare perfumes!

If you were to actually stop and think for a while about these three famous heroes of history, Napoleon, Charlemagne, and Alexander the Great, what descriptive adjectives would you use to capture their personalities and characteristics?

No doubt, you would call them ambitious, daring, courageous, and virile, but would you believe these very men were all passionately fond of perfume and used it in lavish quantities?

Yes, it is true! We are told Napoleon doused himself with at least sixty quarts of perfume and cologne every month. His favorite, remember, was the fragrance of violet.

It is said the court of Charlemagne constantly smelled like a rose garden.

While Alexander the Great had his tunics soaked with aromatic perfumes, his floors were sprinkled with perfumes, and all during his life he was surrounded by fragrant resins, while myrrh was burned for his pleasure.

In days gone by, it was not considered effeminate for men to use fragrances of all kinds and types. Quite the contrary, it was a mark of great culture and the greatest refinement to do so.

For many years a full-blooded American male would have

thoroughly thrashed anyone daring to ask him what cologne he was wearing. But today, it is quite different. Such a query becomes a compliment to a man's good taste in perfect grooming!

At this point, I wish I dared to include in my little book excerpts from letters in my files which I have received from many famous people of today. But they have been written to me in confidence.

Upon occasions, when I have needed a bit of courage, a lift in spirits, I have re-read these letters.

As I have read how my precious little perfumes have brought joy and happiness to so many famous people, it has given me a wonderful, warm glow deep down inside.

It has convinced me the many hours I have spent with my hobby, both in research and experimentation, have not been spent in vain.

My cup runneth over with thankfulness as I realize how much the study of perfume has done for me. To share this great joy with others, knowing the fruits of my effort have made life just a tiny bit happier for them, makes me realize more and more that "Perfume has the charm to calm the troubled heart." And also to make life much happier in its passing.

Chapter 21

The Attitudes Of Perfume

The attitudes of perfume are many.

Research in the field of perfume proves the study of fragrance to be an outline of both science and history. It is an excursion into the realm of beauty and the great mysteries of the ancients.

The story also involves the chemistry of the ancients. The study of their chemistry reaffirms the statement that, "there is nothing new under the sun." For, with their very crude equipment, they accomplished a perfection in perfume entirely unkown in the industry today.

For _they_ discovered the emotional value of perfume!

In their simple way, they came to realize the emotions could be as easily, and much more completely, through the sense of smell by perfumes than through the sense of taste by strong liquors or "fermenti."

The study of chemistry in perfume involves molecules so tiny they would be invisible under the most powerful electron microscope, even if they were magnified a million fold. It is the story of the most painstaking chemical research to unlock the mysteries surrounding the structure of oil molecules and their enzyme chain reactions.

Yet, with all this research and knowledge as a background, odor defies all chemical explanation and scientific classification, thus remaining veiled by the mysteries of the unkown in many ways.

Whether they realized it or not, the men of ancient times, who first made perfumes, were constantly seeking to distill something resembling the very essence of human romance; something which would convey in every drop the glorious drama of true living to the greatest extent. They desired to know the intimations of the soul and the spiritual thrill of true love.

The power of perfume is most significant. It invests with charm and distinction everything with which it ever comes in contact. Poets tell us its lingering touch is often thought of as that of a beautiful woman. For that reason, in many languages perfume is expressed in the feminine gender.

Perfumers know to work or trade in perfume is to actually give your life, and your very soul, over to romance!

It is said, regardless of the beauty or value of jewels, the jeweler who sells them is never thought to be touched with any particular degree of romance. They are merely considered merchants. But with those who make or sell perfume, the association is quite different! It has been written, "there is nothing that men have sold that so dignifies the merchant!"

We do not think of perfumers as merchants, really. Instead, we think of them as dealers in dreams.

The scope of perfumery is extremely broad. The perfumer, before he dares to venture forth into this field, must be familiar with the science of botany, chemistry, and especially organic chemistry. Endless research is involved when preparing for this field. The study leads to countless problems of anthropology, evolution, mythology, religion, physiology and even medical therapy. The reason? Perfume has a history as old as time itself.

Babylonians and Assyrians, 3,000 years ago we are told, believed burning aromatic plants which have odors disliked by evil spirits would restore the sick to health by clearing the air of demons which they thought carried and caused disease.

For many centuries people delighted in aromatic woods and plants simply because the fragrance revived their spirits. We read in the works of John Parkinson (1629) the following, "Many herbes and flowers with their/ fragrant sweet smells doe comfort, and/ as it were revive the spirits and perfume/ the whole house."

Still in later years, a few cold blooded scientists patiently explained that this delight was really due to the curative effects of which the ancient ones had been instinctively but unscientifically aware!

Maybe so...but personally, I find it much more fun to be romantic than scientific where perfume is concerned.

I still believe "when a box of essences is broken on the air" the mind forgets its cares and the soul is set free to dream.

My reward for this belief has been more than generous.

My renumeration has been the precious peace of mind which comes to me as I deeply inhale each blessed fragrance and listen as it softly whispers to me its secret and endless promise of mystery, love and romance!

Perfume can lift you out of the rut of mediocrity and place you on those aspiring heights where true freedom and happiness rule supreme.

It is said of all the means of perfect expressiveness of woman, none can be more delicate or more innocent than perfume.

She is able by means of fragrance to eliminate all inhibitions, and all complexes which act as shackles and free her very soul. Many are able to express themselves with perfume, but were they to depend on their art with words, they would be forever bound in silence.

A woman is able to express her innermost thoughts, and her deeply hidden desires and lovliest dreams through the fragrant medium of perfume; if only she understands its enchanting language.

Once in the lifetime of every lady she will discover the perfume fragrance which is truly hers and hers alone. From this embryo she will build other combinations, variations to suit her moods, but deep within her she will always remember the one fragrance which sets her free and gives her dreams of another world.

Looking back a bit, we are told the Crusaders were largely

responsible for the importation of perfume into Europe. They had been greatly impressed during their voyages with the refinement and gentile manner of those who enjoyed the use of perfume in their daily lives, and attitudes of luxury.

Consequently, they brought perfume into Europe not only as a luxury, but to better civilize those living at that time as well.

One of the greatest accomplishments of those Knights of the Cross was to introduce rose water into the fingerbowls of guests at great banquets and dinners of state. Believe me, this must have done much to improve the table manners of Europe!

They ate much of their food, especially the meat, with their fingers. So the addition of the rose water bath for their fingertips must have been a tremendous improvement. Thus, through perfume, they were imbued with a sense of cleanliness.

Then, too, there once was a time, in the middle ages, when it was scarcely safe for a man to place in writing even a love note to his lady-love. It was then that even strong and courageous men learned to let the perfume of flowers carry forth for them their messages of love and desire.

A small fragrant bouquet in the 17th century was called a "tussie-mussie," or nosegay, which was often sent not only for a "sight and smell" by a lover, but also for a message to his true love.

This was a cluster of flowers with the center one being

the most important in sentiment as well as in fragrance. The message this was intended to carry was emphasized by surrounding flowers and fragrant leaves.

The bouquet was placed in an elaborate filigree holder usually surrounded with a bit of delicate lace for those who could afford such luxurious love-making.

As an example of this language of love in perfume, I pass on to you the following quote simply as a guide: dwarf sunflowers--your devout adorer; Austrian rose--thou art all that is lovely; scarlet ranunculus--I am dazzled by your charm; pelargonium quercifolium--lady, deign to smile at me; red tulip--declaration of love; jonquille--I desire a return of affection; Queen's rocket--you are the Queen of coquettes; lady's slipper--capricious beauty; and to the above declaration of love or admiration, the lady could answer with a single flower as follows. Spiderwart--I esteem but do not love you; yellow rose--the decrease of love on better acquaintance; dogwood--indifference; and china aster--I will think of it.

If the lady wished to positively dash all his hopes, reject him, or simply wished to tell him "Get Lost!" she sent him a variegated pink or an iceplant blossom, and that was quite sufficient!

By sending him a single yellow hyacinth, she conveyed the message, "The heart demands other incense than flattery." Just imagine that!

All this seems a bit strange to us in this day and age, but in those days, so long ago, it gave them an attitude of expressing themselves in security, the security of perfume.

Odor is really the story of language. It was, without a doubt, man's first effort of expression emotionally, and has remained down through the centuries a surpassing medium for such expression. In fact, men express their love even today with gifts of perfume to their ladies. Sales records prove men purchase more perfume today than women. Right here in America more than 95% of the perfume today is purchased by men, bless their hearts!

From one very old book, we learn there were some in ancient times who could not possibly enjoy sniffing unless they were eating something at the same time! The first perfumers even learned how to surmount this difficulty.

We find a recipe for an ancient delicacy, "Crystallized Rose Petals," described as lovely to look at and most delicious to taste. It reads, "Gather petals in the morning, wash in cold water, dry carefully./ Beat the white of an egg, add a scant teaspoon of water. Dip/ each petal in mixture then in sugar. Let dry and serve."

This was a favorite confection, especially after a heavy meal. Sounds delicious, doesn't it? Certainly it must have created an attitude of complete satisfaction.

Perfume chemists are now busy making house paints which

smell like lilac (for those allergic to the odor of regular paint), mousetraps which smell like cheese, and wrapping paper which bears to us the enticing fragrance of good food for the purpose of stimulating the appetite. What next?

One of my importers brought me a priceless story, which he swears is true. It seems a man, who lives in his home town, has made a fortune by perfecting a spray which used car salesmen employ to spray the insides of used cars. The used car is immediately imbued with the fragrance of a brand new car just off of the assembly line!

Only recently, I have learned, when deer ran wild on roads and runways at Vanderberg Air Force Base in California, they attached rags which had been soaked in canned "lion scent" to stakes driven into the ground and got rid of the deer in short order!

Animals are most susceptible to odors.

Trappers know a mixture of musk and oil of rhodium is positively irresistable to bears. Therefore, they use this fragrance to bait their traps.

Every professional trapper, it seems, has his own formulas for making scents which lure wild animals to his traps. Some are pleasant, some are vile, but all serve their purpose to perfection.

Just imagine, to an Eskimo the odor of rancid fat is the most delectable fragrance in all the world!

Some women of the Near East, it is said, simply revel in the rank odor of stale perspiration of the opposite sex. They go so far in this respect it is almost unbelievable. Some "perfume" their hankerchiefs by having their gentlemen friends (rarely husbands) tuck them under their armpits when engaged in manual labor and perspiring freely!

Some tribes, such as the Warrau Indians of the coastal swamps of the Orinoco Delta, prefer the odor of dead fish to all other scents and often rub their bodies, their faces and even their hair with the slime and oil from the fish they catch.

My motto has always been, "To each his own!" But this is really carrying things a little too far, isn't it?

A famous trader found, quite by accident, while on an expedition in South America that the Aborigines were greatly attracted to the odor of onion peels. When he used a garlic bud as a means of stimulating their interest, the bartering became really hectic. Later on he learned they liked the scent, it was true, but they had discovered the smell of onions or garlic actually discouraged the attention of the fleas which were eating them alive in the village!

Many insects have decided likes and dislikes regarding scents. In our southern and middle states, the odor of osage orange is most obnoxious to wood-ticks and spiders.

Clothes moths abhor the odor of red cedar. The scent of castor oil bean plants, or citronella, will usually keep mosquitos at a respectful distance.

The oil of the West Indian Crabtree is hated by redbugs or jiggers.

The fragrance of catnip greatly pleases the house cat, and causes him to jump with glee. Lavender has the same effect upon other members of the cat family, such as lions and tigers.

The wildest birds can be easily tamed by rubbing the oil of bergamot on their nostrils. The problem, or course, would be catching the bird to do the rubbing on the nostrils. Seems a bit difficult--and slightly silly-to me.

Recalling the stories of this chapter, I believe you will realize this: down through the ages, perfume has been used to discourage evil spirits and pests, but its main purpose has been to create an attitude of true happiness for all mankind.

Chapter 22

The Art Of Wearing Perfume
To Advantage

It is said a quite plain and retiring woman wearing a single diamond, in good taste and simple grace, may appear as lovely as a Princess in the eyes of a man. Yet, another woman may be loaded down with diamonds and only appear vulgar and revolting to men. This reaction also applies to the use of perfume. Why? Because many women select a perfume far beyond their ability to fulfill its promise!

As you know, most of our fads and fashions are brought to us from France. Few of us stop to realize, however, that in France, as in every other country, there are two kinds of women: French Ladies, and those who are not quite so lady-like. Unfortunately, it is the latter type who set most of our fads and fashions for us, and we blindly follow.

I warn you this is not wise when it comes to perfume, and I shall tell you why!

French Ladies spend hours on selecting their perfumes. They would never think of wearing a perfume simply because a friend wore it well, or because it was highly advertised on television, or because their favorite movie actress wore it to advantage. Indeed not!

A French Lady requires a perfume that actually <u>does</u> something for her, and would never consider anything less.

The French have always believed a modiste is able to design a gown just for you and you alone; a milliner may create a chapeau to bring forth your hidden beauty; but only a perfumer is able to create for you a mood.

A French Lady asks for only three things of this life. She asks just that she be admired, that she be loved, and that she be remembered. She knows perfume will accomplish all three for her and even more.

When she attends a luncheon, a business meeting, or to enjoy a day of shopping, she applies her perfume, which she has so carefully selected, to the pulses of her wrists. She knows it requires heat and action to prompt perfume to accomplish its purpose.

As she talks, she gestures constantly, and with each gesture the blood rushes to her fingertips causing the pulses at the wrists to quicken their pace. Those, to whom she is speaking, are enthralled by each burst of fragrance. As they enjoy the delightful aroma, they realize how diligent her search must have been, how many hours it must have taken her to find exactly the right perfume for her. They admire her for her selection, and thus the first requirment of her life has been fulfulled!

When a French Lady desires to be loved, she can recall the custom of the ancient Persian women and place her perfume on her breasts, where the beating of her heart, the rising and falling of her breasts with every breath, and the heat of her bosom brings forth wave upon wave of the heavenly fragrance she has selected, with great care for that special occasion, and she is loved.

When a French Lady seeks to be remembered, perfume also plays a crucial role. She places her perfume, very carefully, on the pulses of her knees! As she walks, dances or even just leaves the room, the swishing of her skirts and the warmth of her knees constantly revive the fragrance. It leaves a very distinct but delightfully scented aura trailing behind her. She is remembered for her perfume...if for no other reason!

However, it is well to remember, when a French woman, who is not quite so lady-like wears her perfume she applies it behind her ears, especially when she is about to make a conquest. Do be careful!

Often I am asked, "Do you have a perfume that will do something for me?" Of course I have, but the question is, 'What do you want to do?'"

Perfume is the most concentrated form of fragrance, and should be worn directly on the pulse spots of the body, such as the throat, wrists, and crook of the elbow, places where heat is generated. Warmth and action causes real perfume to live again and again in endless cycles.

A drop or two of fragrance may also be applied to the shoulders (for dancing), or on the upper lip (when desirous of being kissed) but always a drop or two should be applied in the cleft of the bosom for, as the heat of the body and the beating of the heart activates those few drops, the dividends are tremendous! The lady enjoys the fragrance as it is mixed and wafted up towards her face and all those around her are entranced by its ever changing attitude.

It is positively wasteful to apply perfume to the ears, back of the ears or at the nape of the neck because then only the people walking behind you can enjoy it. This last statement brings to mind a brand new thought. You know, there is the type who enjoys being followed! For these, the incorrect application of perfume might save the wear and tear of droping hankerchiefs!

When the perfume, which is absolutely right for you, is applied properly it will work wonders. Men always notice a lady's perfume long before they notice her dress or hairdo. Why? Because it is the only natural instinct for a man to be attracted, or repulsed, by odors. All the civilization of centuries yet to come can never hope to quell this animal instinct upon which he relies so greatly as a guide.

Many times a woman will come to me with this problem: "I simply adore this perfume," she will say, "it smells heavenly on my sister and several of my friends, but on me it smells just horrible! What is the matter with it?"

Should this ever happen to you, please do not blame the the perfume! The answer is quite simple. The perfume in question simply is not compatible with your particular body chemistry.

Many people have oily skin, especially when the sebum has a strong odor of its own, find it impossible to wear perfume, cologne, or toilet water which contains alcohol. It takes on a "sour," almost rancid odor within only a few minutes. A body powder, after a thorough bath with felsnaptha soap, is the solution. I have suggested this to hundreds--each one has returned to thank me.

As I have previously said, Mary Lynne's perfume is made as the ancients made their perfumes long ago. No alcohol is used.

My secret is this: I have learned how to collect the enzymes from the plants, the resins, the flowers and the animals, and then hold them alive in suspended animation. My perfumes will never evaporate; they will go to sleep, though, it allowed to stand perfectly still for any length of time. For that reason I have always instructed those who wear my perfume to first awaken them. How?

Before applying them to the skin, simply grab the little bottle firmly in your right hand and smartly strike the bottom of the bottle on the heel of the left hand. Do this until tiny bubles appear in the bottle (just like ginger ale bubbles). Then you will know the perfume is awakened and ready to go to work for you.

As you remove the cap from the bottle, the fragrance rushes up out of the tiny dropper top. As this scent combines with the oxygen and hydrogen of the air, the first enzyme chain reaction takes place.

What is an "enzyme chain reaction?" I can only explain what it means to me. As I think of an enzyme chain reaction, I recall a long string of tiny red firecrackers such as we used to have on the Fourth of July, when we were children. When you lit the fuse, first one firecracker would explode, then the fire would quickly ignite the next one, until a rippling chain of exploding firecrackers went, "bang,bang, bang!" one right after another.

So it is with the enzyme chain reaction. When the enzymes of your body chemistry unite with the enzymes of the perfume, it makes a very tiny explosion of the precious little perfume cells, reviving the fragrance again and again for hours. All this is accomplished merely by body heat and activity. You cannot hear the tiny explosions, but you certainly are able to enjoy their fragrance, one after the other!

As you touch your wrist (on the pulse) with the little spill-proof opening of the perfume bottle, the heat of your flesh brings about the second enzyme chain reaction.

Then rub your two wrists together causing a light friction. This brings about the third enzyme chain reaction. Warmth and action really start the enzymes into motion, and from then on the perfume becomes a part of you. Your own body chemistry takes over and the following enzyme chain reactions are caused by your body chemistry reactions upon the perfume.

Should we, as an experiment, apply the perfume to our skin, if your skin is moist and my skin is dry, we will experience two entirely different enzyme chain reactions. As perspiration is heavy in ammonia, the combination with the perfume gives you a very sharp effect and a tangy odor. A dry skin absoorbs the perfume quickly and the resulting odor is mellow and inclined to be sweeter, softer and more elusive.

If your skin is dry but my skin is oily then another entirely different enzyme chain reaction results. Again the dry skin absorbs the perfume, mellowing as it holds it tightly. The fragrance is soft and pleasing. However, the sebum, which is the oil secreted by the sebaceous glands, is made up almost entirely of the natural enzymes of the body. Therefore, when the perfume unites with all those many enzymes, they will really have a hey-day!

The enzymes of the sebum together with the enzymes of the perfume activate into one grand and glorious jam-session. The person with an oily skin really gets her money's worth when she uses my perfume. The enzyme chain reactions just go on and on for hours, bringing about new and wonderful changes constantly.

If your body chemistry is on the acid side, then the perfume will give you a tart fragrant reaction. If your body is on the alkaline side, the perfume on you will be hot, firey and full of life.

However, if you retain the toxic poisons of your body,

and I burn up my toxins quickly, the enzyme chain reaction, as we apply the perfume to our skin, is again entirely different!

The person who retains the toxins has the most difficult time of all selecting a pleasing perfume. It takes genuine art for a perfumer to completely cover the toxic odor.(that of something dead or decayed from within) with a pleasing fragrance. Such a person has to wear a heavier, more exotic perfume to overpower the unpleasant odor of their body.

The person, who has perfect body eliminations of all the toxins is the perfumers pride and joy. They are the ones who can wear the lightest of fragrances, even bordering upon the ethereal, and everyone exclaims how heavenly they wear their perfume!

I have a favorite example of the powerful enzyme chain reactions which I frequently encounter and use in my lectures.

As, I think I have told you previously, I always demonstrate my lectures with sniffs. When we get to the "Perfumes of Persia," I ask everyone present to find a secluded spot somewhere way up on their arm where they can apply just one drop of my "Persian Nights" perfume. I simply instruct them to rub it briskly with the back of their hand and then let it rest until the end of my lecture, when I will have a secret to tell them.

All during my lecture, the enzyme chain reactions are working like mad where each one has applied a drop of "Persian

Nights" perfume. By the end of my lecture, it is really ready!

At the very last, I always say, "And now for the secret of 'Persian Nights' I want you to find the secluded spot where you applied this precious perfume on yourselves...sniff it. Then sniff the "Persian Nights" on the neighbor sitting beside you. Not two of them will be the same, in spite of the fact that you are all wearing the same perfume. Why? Because there are 59 different perfume oils--all loaded with enzymes--in "Persian Nights." Each of you have at least seven enzyme chain reactions hidden in your body. As a result, there are almost 500 enzyme chain reactions involved which makes "Persian Nights" your very own individual perfume.

I find the ancient physicians often used the different scents applied to the skin of patients in making a diagnosis. This leads me to believe the ancients knew far more about the enzyme chain reactions of our bodies than we know today, even with our higher education.

When I endeavor to help a lady select a perfume for herself, I always suggest she sit at my sniffing table and quietly sniff first one bottle and then another. Always there will be several she will like better than the others. I tell her the story of each of those she found pleasing, then I have her try one then another on her skin. As we sit and chat, the enzyme chain reactions are taking place. After a few minutes, I then suggest that she sniff where she applied the perfume, and I sniff it too. It is quite simple to find the one made just for her!

As she sniffs one she will say, "Lovely, isn't it?" or "How nice." But the perfume which causes her to exclaim, "Oh! This is the one!" with stars in her eyes is the one I suggest that she select for her own.

Quite often there will be several which cause them to exclaim such remarks and really "sends" them. If they can afford to buy them all, it is wonderful, but often we must decide on only one or two for that visit. It is then that I suggest one of the lighter fragrances for daytime wear and one of the heavier fragrances for nighttime wear. We always carefully note the others, however, and soon they return to claim the rest.

When a person comes to me for help in selecting a perfume for a gift it is a bit more difficult. I always ask a few questions about the person to whom the perfume is to be given. Their coloring? Their favorite flower? Their favorite perfume or commercial cologne? Their type in general?

From these answers, I form a mental picture of the person who is to be pleased with my perfume, and work from there.

Seldom do I select one of the heavier perfumes for a person has never before worn my perfume. I have found it wise to start with a lighter fragrance. "Fantasy" makes a delightful gift. As it contains 753 different fragrances all blended into one perfume; it is sure to please. Then, too, as it is worn, it changes almost on the hour, projecting many different moods in a single evening!

We are told there are really only "Seven Basic Fragrances." These represent and project seven quite different moods.

The "Single Floral" is for the person who desires the simple life, uncluttered, uncrowded, peaceful and quiet. Invariably, that person has one, perhaps two, flowers which are her favorites. By surrounding herself with the fragrance of that one flower, one at a time, she can assume the attitude of rose for love, lilac for a feeling of spring, and jasmine for a dreamy, and gay, evening.

While "Floral Bouquet" becomes a bit more complicated. Here a variety of moods, as well as fragrances, are blended together. Those who enjoy social groups for more than solitude, I find, prefer a blend. This type of fragrance definitely attracts more interest and compliments from others. Blondes thrive on these!

"Spicy" perfumes are the type worn to make one more alert. Usually they are pungent, heavy, and more frequently worn by a dark flashing person. Commercial perfumes are often made from the bark of the spice, but I find the blossoms of the spice are far more soft and gentle to the nose and to the sense of smell in general. Cinnamon, clove and ginger predominate but are complimented beautifully by the addition of a fragrance or two extracted from some of the spicy flowers, such as the

Old Fashioned Pink. This adds a warmth most attractive.

The "Woodsy-Mossy" fragrance is considered a type, however, that is really two types in one. There is the "woody" type which can only be portrayed by the oils squeezed from wood. This is for the person who loves the smell of sawdust, newly cut wood and wooden chests. Then there is person who adores the fragrance found only in the forest. This is the scent of trees, especially the evergreen, cedar and walnut with the the undertone of moss covered stones beside a bubbling brook. This represents the great outdoors and all its glorious freshness.

Then there is the "Oriental" type of perfume. This attracts the attention of those who desire to capture the subtleties and mysteries of the orient. Usually a person who enjoys oriental art finds this type of perfume to be most interesting. The oriental scents are always well fortified with the animal fixatives for, after all, it is the musk, civet and ambergris that creates the desire in the first place.

A "Fruity" perfume is characteristic of a very special type, for it conveys the fragrance of citrus fruits. For this type, I employ the oils extracted from the blossoms of the lemon, lime and sweet orange. These are simply softer and more effeminate. The oil extracted from the rind of the fruit is more tart, more "nippy" and much more dashing, therefore used more frequently for the men. Often men, who have just sniffed my "Citrus" for the first time will remark, "Wow! This is terrific! Do we wear it or drink it?"

In the commercial perfume world, the seventh type is known as the "Modern Blend" because they are aldehydes, chemicals and synthetics made to resemble true fragrances. They have become extremely successful, but I am sorry to say they leave me cold.

Therefore, for my seventh type, I list "Grasses." There are those who dearly love the fragrances which can only be found at twilight, at the appearance of the first evening star, in the great outdoors. There is a dewy effervescense at a magic moment like that, which we all yearn to have and to hold forever as our own. This fragrance of escape, from the problems of life, is most difficult to capture. I tried desperately to do so when blending my "Evening Star." Many tell me I have been successful. To me, it is a dream fulfilled.

The ancients appreciated the emotional value of perfume far more than we realize today.

Once upon a time there was a wealthy king. So wealthy he could well afford to have a very special perfume made exclusively for his own pleasure. This particular perfume was most precious to him, not for its monetary value alone, but especially for its emotional value to him.

Whenever he inhaled deep sniffs of its fragrance, he found it gave hime exceptional strength and courage beyond all comparison. This was invaluable to him as he had great plans for his future.

Just before entering the most important battle of his entire career, he felt the need of great courage as never before. Consequently, he literally soaked himself in his precious perfume.

Once again, it did wonders for him. He entered battle full of courage and confidence, but things did not go as he planned. It was finally necessary for him to take flight, and he barely escaped from the battlefield with his life!

Before the smoke of battle cleared, he managed to take refuge within a cave he discovered in the hillside far above the field where the fighting had taken place.

It was his misfortune that during the night there came a gentle rain.

Suddenly, as the cool damp breeze wafted towards the enemy chieftan, he sprang to his feet, summoned his trusty warriors to accompany him, and ran towards the hills!

They went directly to the cave sheltering the king, and were led there by the scent of his perfume. The fragrance of which had been held close to the earth by the sudden humidity. The perfume which had given the king such courage also led to his own capture and death.

This proves what good for us can also be bad for us...when used in excess.

Therefore, always remember: perfume is to pleas, excite, interest, attract, maybe inveigle...but never to choke!

When applying perfume, remember the suggested areas, but be certain you are ready, willing and able to adhere to the consequences!

Where do men wear their perfume? I always instruct them--after a shave or shower--to place one drop of perfume on their wishbone (their chests)--then all their wishes will come true!

Don't you believe this is possible?

Well, please don't knock it until you have tried it! It could work wonders for you!

Chapter 23

Sniffing Can Be Fun

These days, I am often reminded of the story of the ancient Jews.

When the Jews finally emerged from their many years of slavery under the Egyptians, they were greatly depressed. The feeling of oppression they knew was to the point of being overpowering.

So what did they do?

Realizing Man's extremity is God's opportunity, they prayed.

In answer to their prayers, God gave Moses the formula for making perfume.

In those days there were no chemicals, no synthetics so, under the instruction of God, Moses took plants, a natural source, from which to make the first perfume.

What was the result?

As the Jews prayed, they inhaled the fragrance which made them calm once more; it gave them courage, and soon they became better able to cope with life. But, most valuable of all, the fragrance they inhaled <u>lifted them up</u> from their feelings of depression, and oppression, allowing them to experience an extension of thought and an expansion of the mind. This was exactly what they needed to give them the courage to go on to bigger and better things awaiting them in the future.

Mind you, all this was accomplished without a "hang-over,"

or any other detrimental after effects whatever--either mentally, emotionally or physically--simply because the fragrance inhaled was made from natural sources: God's own living plants!

When I read of the "glue-sniffing" today, as well as the craze for drugs among our youth of this particular period, I realize more and more that they, too, are searching for deliverance into a promised land. They, quite unconsciously, are depending upon their olfactory nerve, their sense of smell, for guidance. They are on the right track--but the wrong train!

Unfortunately today we live in a "Chemical Age." An "Age of Synthetics" which we are schooled to believe are just as good, or even better, as the natural things in life.

This is definitely not true. How do I know?

Let us retrogress a bit.

The days and years, which followed my discovery of the formula in the Bible, were extremely hectic.

The Depression was still with us. Finally the Depression ended. But, before we could regain our equilibrium, we were suddenly involved in an all consuming war!

During those years, it was necessary for me to work long hours endeavoring to help the hundreds who came to us seeking assistance, because they were almost beside themselves with worry or concern. I opened the office and 8:00 a.m. and often said "Good Night" to my last patient at 11:30 p.m.

Fear, consernation, stark horror and frustration were radiated by everyone in those times. My days were so filled with trying to separate the real from the unreal, that by night I was exhausted.

As each patient left, their gratitude shone, often with tears, in their eyes. But, after the door closed, even though my heart was thankful for the gift God had given me enabling me to help so many, my body was exhausted and my mind numb. How many times I found myself wondering if I would ever make it home!

It was <u>then</u>, on such a sleepless night, I reached out for the little Bible resting on the bedside table and, as I prayed for guidance, the Bible opened in my hands to Exodus 30:34. As I read, in heavy black print, the word "perfume" appeared at the top of the page, and I was immediately alert. My prayers had been answered, for from that moment on my interest has never waned.

Yes, from that night on, I lived through the trials of the days with a gracious understanding I had never known before. There was now a purpose for me to get through each day. My reward was the few moments I could spend with my beloved perfume research in my tiny lab which, then, was only a corner of the utility room in our basement.

That little corner was my escape for many years. Only God, the perfume and I knew the joy it held.

Hourly I became aware of a new reassurance I had never felt before. Mentally I became more alert than at any time in my life. Physically I soon became rugged, positively indestructible!

I soon realized this change was not just my imagination. Others, many in my field, worn to a frazzle with the trials of life, simply marvelled at my endurance and my fortitude. My enthusiasm for life was born anew!

It was then I realized I must share my perfumes with others, so that they, too, could know this great feeling of being uplifted, and encouraged to go on to accomplish the many things awaiting them in their particular future.

These gifts of love, placed where most needed, and watching the forthcoming reactions have been far more rewarding than all the money in the world!

The recipients of those gifts have written letters, through the years, which have filled my files to overflowing. Each and every one describes a personal experience, a hurdle in their life, which has been taken better in stride simply through the use of my fragrance. What greater reward could one ever expect?

The sniffing table in my reception room for years furnished endless, quiet entertainment for my patients as they waited. The little "Sniffers" passed among the guests at my lectures have become internationally famous! The sniffing tray in my lab is, without a doubt, the most popular place in this area, for those who love perfume and wish to try the many fragrances I have designed have come to the right place!

What does all this prove?

To me, it proves conclusively: Sniffing _can_ be fun!

Chapter 23

Much Stranger than Fiction

During my years of perfume research, as well as lecturing from coast to coast, I have received countless "gems" pertaining to perfume, scents, and fragrances from all sorts of the most interesting people.

These wonderful people, it seems, have either heard my lecture, talked with someone who has heard it, have read about my hobby, heard a radio program, or watched me make perfume on television. At any rate, they are always enthused to the hilt as they write!

The magic, as far as I am concerned, is this: in their minds, perfume has been linked with me and my name. What I have managed to do, in my feeble little way, has so tremendously interested them, they have taken time out from their busy days to sit down and either write or call me. You have no idea how happy this always makes me.

My perfume file is filled with letters and notations of their calls. At this time I feel duty bound to mention at least a few of these priceless gems in my book. I mention them in this chapter because I think you will agree with me when I say it has given all concerned a new attitude of happiness. Perfume has given them something new, different, and wonderful to think about...all of it much stranger than fiction!

Thank you with all my heart to those who have contributed

to this most interesting chapter!

One dear old doctor, with whom we worked for many years, wrote, "The word 'oil' itself symbolizes "Spirit" or "Illumination."

"Happiness is a perfume you cannot pour on others without getting a few drops on yourself" was sent by a dear old lady.

'Twas a sultry evening in late summer. The setting? An exclusive country club frequented by the jet set.

Suddenly he was aware of a stately Goddess floating past. The aura of fragrance surrounding her enveloped him completely.

With a start, he realized this was the moment!

This was the fragrance for which he had been awaiting all the long years through.

She turned. Their eyes met. Suddenly he asked, "A thousand pardons...your fragrance? Your perfume? Please, what is it?"

"Happy New Year!"

"Thank you! The same to you! But <u>please</u> your perfume? What is it? Who makes it?"

"The name is 'Happy New Year' by Mary Lynne. She is a dear friend of mine." And she was gone.

After a round of golf, as a guest of a friend, a man was changing in the locker room. The stranger next to him quickly applied a bit of after shave lotion from a simple little bottle kept in his locker.

That fragrance! By far the most mannish scent he had ever encountered. He finally asked its name in a casual sort of way. The reply?

"'Tack Room' for men, made by Mary Lynne. Sorry, old man, it is not sold on the market, you know."

Chapter 25

The Ending......

At last, we have reached the end of our fragrant journey.

Writing this book has been the most tremendous task I have ever undertaken. Yet, I have enjoyed each moment I have spent placing my research, my many experiments, my innermost thoughts and all that goes with it on paper, because I have written it just for you.

The greatest joy of all, I believe, is in sharing the good things of this life. It has made me extremely happy to share this World of Fragrance with you, and which I have had the joy of rediscovering.

If, in this book, I have made you just one tiny bit more perfume conscious, my mission in life has been completed.

Many years ago, I read the works of a great historian who wrote, "The history of the entire world has been written in three fluids: blood, tears and perfume."

During our short lives upon this earth, I feel positive we have seen far too much bloodshed, and cried far too many tears. So, I feel within my heart, it must be time to begin writing history again with perfume!

During my lectures, I closely watch the expression of the faces before me. It is amazing to note, while joining me in the World of Fragrance, not a single one recalls for a moment a care or a worry they have left at home.

Why?

It is because the World of Fragrance is far different from the world in which we live today. There, fact and fancy link hands for one purpose, and one purpose only, and that is simply to have fun!

Most business and professional women dread the time of their retirement. They realize they will then be "over the hill" and "no longer needed."

Can you possibly imagine the great joy that fills my heart to know, in my retirement, I am able--in this war-torn world--to bring one hour of complete happiness to so many hundreds of people each and every week of my life?

Surely my cup of joy runneth over and I am thankful with all my heart.

                              The End

Printed in the United Kingdom
by Lightning Source UK Ltd.
111887UKS00001B/9